How Round Is Your Circle?

How Round Is Your Circle?

WHERE ENGINEERING AND MATHEMATICS MEET

John Bryant and Chris Sangwin

PRINCETON UNIVERSITY PRESS

PRINCETON AND OXFORD

Copyright © 2008 by Princeton University Press

Published by Princeton University Press,
41 William Street, Princeton, New Jersey 08540

In the United Kingdom: Princeton University Press,
3 Market Place, Woodstock, Oxfordshire OX20 1SY

Library of Congress Cataloging-in-Publication Data

Bryant, John, 1934 July 19–
How round is your circle? : where engineering and mathematics meet /
John Bryant and Chris Sangwin.
p. cm.
Includes bibliographical references and index.
ISBN 978-0-691-13118-4 (cloth : alk. paper)
1. Engineering mathematics. 2. Geometry, Plane.
3. Geometry, Algebraic. 4. Geometrical models.
I. Sangwin, C. J. (Christopher J.) II. Title.
TA330.B79 2007
516'.15–dc22   2007032801

British Library Cataloguing-in-Publication Data

A catalogue record for this book is available from the British Library

This book has been composed in Lucida

Typeset by T&T Productions Ltd, London

Printed on acid-free paper ∞

press.princeton.edu

Printed in the United States of America

3 5 7 9 10 8 6 4 2

To Jack and Ralph

To Jon

There are very few things which we know, which are not capable of being reduc'd to a Mathematical Reasoning; and when they cannot it's a sign our knowledge of them is very small and confus'd; and when a Mathematical Reasoning can be had it's as great a folly to make use of any other, as to grope for a thing in the dark, when you have a Candle standing by you.

John Arbuthnot (1692), *Of the Laws of Chance*

Contents

Preface

Mathematics and engineering

W. M. Fletcher (tutor) conceived an idea (c. 1912) that Engineering students should be taught some 'real' mathematics by the *mathematical staff*—'contact with great minds'. The hardworked, hardboiled, and lazy devils hated it as much as I did, to whom, as junior, all dirty work then fell. I asked F. J. Dykes (the sole Lecturer in Engineering) what he would like me to select; all he said was 'Give the buggers plenty of slide-rule'.

Littlewood (1986, p. 142)

The research mathematician, concerned with purely abstract realms, and the mechanical engineer, who grapples with the practical difficulties of the physical world, are unlikely companions. This book is our attempt to illustrate why mathematicians should take the practical problems of engineering seriously for the real mathematical challenges they present. We encourage them to leave their comfortable world of fictitious thin lines, perfect circles and exact numbers, at least occasionally. To the engineer these simply do not exist, and yet as we shall see they cannot be ignored completely. Furthermore, in order to even perform apparently 'trivial' tasks, the engineer owes a lot to mathematics. Moreover, this mathematics is far from the dull repetitive practice of routine tasks which is at least one interpretation of Dyke's 'plenty of slide-rule'.

Mathematicians and engineers are disinclined to agree about anything in public: should the area of a circle be described using the neat formula πr^2 or in terms of the more easily measured diameter as $\frac{1}{4}\pi d^2$, for example? However, there will always be mathematicians who are interested in practical problems and engineers with a serious passion for mathematics. This book, we hope, will be a source of ideas for them and for their students.

In this book we look in some detail at a number of apparently trivial problems, specifically including how to draw a straight line and how to check that something is round. Since circles and straight lines are the simplest geometric curves, these naturally generalize to construction problems for other sorts of curve. There are also associated questions of measurement, which involve constructing lengths and angles. How does one build 'the first' protractor, or, at an even more basic level, divide lengths and angles into halves, thirds or any specified ratio? Is this even possible? And, if it is impossible in the strict mathematical sense, when does one need to care or when can an engineering approximation suffice? Most of the topics we have selected are surprising, others are clearly well beyond what we imagine to be possible. Rollers which are not round, and stacks of dominoes which lean arbitrarily far without falling over, for example. We promise that examining these seemingly pointless questions will reveal some of the hidden complexities of the apparently simple.

In many of the chapters we explain how to build physical models or undertake experiments to illustrate instances of these problems. One practical demonstration even illustrates that the mathematics 'works'. Of course, such an illustration is not a proof as the mathematician knows it. Nevertheless, we believe that making such models and other practical activities are beneficial. This is both from our personal experiences of trying to grasp results that our intuition rejects, and also from working with students and teachers. We also discuss the practical limits associated with both the mathematics and the modelling process. How far may one tilt a stack before it collapses? When does the width of the saw blade matter? These questions inevitably lead to a better understanding of, and appreciation for, the mathematics.

Some experiments may be undertaken in minutes with standard drawing instruments—pencil and paper. Models can be constructed out of commercially available toy sets, or materials that are readily to hand. Others require sophisticated machine tools and workshop facilities. Of course, these latter models are less suitable for classroom construction activities.

We have tried to keep the mathematical explanations intuitive and geometrical, resorting to calculus only when absolutely necessary. Even if the finer points of all the mathematics are not fully explained, we sincerely hope you enjoy the activities and develop a deeper appreciation of some apparently simple topics.

We also wanted to alert you to what were commonplace scientific instruments that have now been rendered obsolete in the recent age of digital technology. The majority of these are analogue devices and most are a cunning mix of the physical and the mathematical. Many are surprising, especially when first encountered. So, before you dismiss them as old hat, we ask that you imagine the reaction of the novice, or the student so immersed in virtual reality that their appreciation of the physical has been impoverished. After all, everything fails to be surprising once it is completely understood! Our experiences of using such instruments to motivate mathematics are very positive indeed. Often such objects (especially slide rules, and occasionally Amsler planimeters) may be obtained second-hand as antiques, sometimes at very little expense. This is currently a practical way for those without workshop facilities or machine tools to obtain a small selection of the best quality mathematical models.

Notes

We suggest practical work with pencil and paper and encourage making models on the kitchen table or in a workshop as ways of bringing more life to mathematics. Our final suggestion is that you go out and see engineering in practice. If we take just one example, the stationary steam engine, we can see one of the straight-line linkages for real in a museum or better still on an original engine preserved *in situ*. In both instances there is the possibility that they will be running, although perhaps not under steam. Models of these can be seen at many model engineering exhibitions. Most towns have their own model engineering societies and they put on annual displays. There are also the larger regional and national exhibitions, where models run on compressed air. It is often easier to understand the working of a linkage in scale model form than it is when looking at a full-sized prototype, where the sheer physical size of these engines makes it difficult to appreciate them properly from a single viewing position. It may need a walk of several metres to move from one end of a beam to the other on some early pumping engines, or a machine might occupy several floors of a building.

Trade stands at exhibitions provide an excellent opportunity to see what is available in terms of plans, raw materials and tools for you to make your own models. The stewards at model engineering society stands will always provide advice and guidance. Two useful sources of information are Hayes (1990) and the fortnightly publication *The Model Engineer*, which gives a calendar of events in each issue.

There have, of course, been many previous books that have explored physically motivated mathematics, or illustrated mathematical results with physical models. Perhaps the most famous is Cundy and Rollet (1961), and we would like to record our gratitude for the encouragement which we received from the late H. Martyn Cundy in preparing this work. Other excellent books are those of Cadwell (1966), which is quite mathematical in its approach, and Bolt and Hiscocks (1970), which contains many activities for school students. Other books, such as the very well-known book of Rouse Ball (1960), also contain much in common with our work, although again the emphasis is on the mathematics, not on the relationships with the physical world.

We see great value in making physical models as mathematical experiments, and have done this with the majority of the examples contained in this book. Indeed, the photographed models come from our own collections. We also recommend experimenting with dynamic geometry and computer algebra packages. There are many commercial systems available, but the dynamic geometry package GeoGebra (www.geogebra.at) and the computer algebra system Maxima (maxima.sourceforge.net) are freely available and were used extensively in producing the illustrations for this book.

Acknowledgements

We would like to thank the numerous colleagues, librarians and museum curators who have, over the years, so generously given their professional advice and encouragement. We would particularly like to thank Professor Achim Ilchmann for recognizing two like minds. We also thank Chris Parker for advice and help in making the solid shown in plate 19.

Dr Bryant writes in particular: I would like to record my thanks to Dr Martin Jenkins for all his enthusiastic support and encouragement; the staff of the mechanical workshops in the School of Engineering at the University of Exeter; my daughter Kate Bryant for all her help with the photography; and most of all my wife Margaret for everlasting patience and tolerance.

How Round Is Your Circle?

Chapter 1

HARD LINES

I have been obliged to confide the greater portion of the theoretical part of the present work to some mathematical assistants, whose algebra has, I fear, sometimes risen to a needless luxuriance, and in whose superfine speculations the engineer may perhaps discern the hand of a tyro.

Bourne (1846)

There are many convincing ways to justify a result. A scientist gathers *evidence* by undertaking a systematic experiment. One can undertake mathematical experiments, such as a sequence of calculations. Another kind of experiment is to draw a picture, be it on paper or sketched in the sand with a stick. Few, if any, mathematicians would now accept a picture as a valid *proof* but sketches do provide us with the simplest and most direct form of *mathematical experiment.* When undertaking such an experiment we ask you to think of it as representing a whole class of similar ones. What can you change without removing the essence of what you are doing? What must stay the same? And then, of course, decide how you can justify this.

So that we might be definite in the difference between a mathematical proof and an illustration, let us begin with an example. This is a theorem from Euclid's *Elements*, book III, part of proposition 31 (Euclid 1956, volume 2, p. 61), which is encountered early in school geometry connected with a circle.

Theorem 1.1. *Take any circle, and any diameter (from A to B, say), and any other point P on the circle. Then the triangle APB is a right-angled triangle, with right angle at P. (This is illustrated in figure 1.1.)*

1

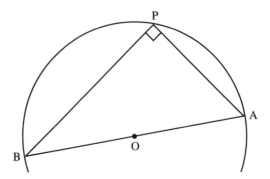

Figure 1.1. Illustrating theorem 1.1.

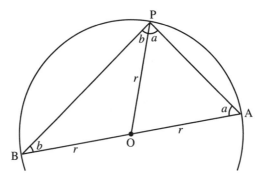

Figure 1.2. Illustrating the proof of theorem 1.1.

Proof. To the diagram add the radius OP. Then OP = OA and OP = OB, so that we have two isosceles triangles AOP and BOP. The base angles of isosceles triangles are equal. Call them a and b, respectively (see figure 1.2). Since the interior angles of any triangle sum to $180°$, we have for our triangle APB that $2a + 2b = 180°$. That is to say $a + b = 90°$. Hence the angle APB is a right angle. ☐

This is a surprising result with a mathematical proof which is beautiful and elegant. It removes any doubt as to the truth of the theorem, but to illustrate and motivate what follows we would like to encourage you to make a paper illustration of this theorem, to confirm the result physically. That is, draw a circle, a diameter and chose another point. Cut out the resulting triangle. How does one confirm that the angle is a right angle? One reliable

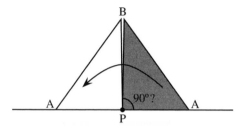

Figure 1.3. Checking that an angle is indeed a right angle.

method is to take a fresh sheet of paper and draw a straight line. Mark a point on this line and match it up with the point P, with the side PA on the line itself. Now mark the position of the point B, and hence the line PB. Flip the triangle out of the plane of the paper and place the side PA onto the line again, but this time on the other side. Now look at the position of B. This will remain unchanged if and only if the angle APB is a right angle. This procedure is reliable in practice and is illustrated in figure 1.3. We can also cut along the radius OP and confirm that the two resulting triangles are isosceles by folding. This also gives us the opportunity to confirm the steps in the proof which assert that the base angles in an isosceles triangle (*a* and *b* in our case) are equal. For our purposes in proving theorem 1.1, it does not matter how many degrees are in a triangle, as long as every triangle has the same quantity of internal angle. This can be physically demonstrated in a final destructive act, by ripping off the corners of a triangle and arranging them in a straight line.

Actually, we need to backtrack a little. While the above procedure confirms that the angle is indeed a right angle, we need to think carefully about some assumptions we have taken for granted.

There is no doubt that a pair of compasses will draw a circle. Since a circle is *defined* to be the set of points a given distance from the centre, the shape you draw with a good pair of compasses is a circle. However, we need to be a little more careful when we draw the diameter and radius. To do this we need to draw a *straight line*. How does one do this? If you were tempted to reach for a ruler, then consider the problem of making a ruler in the first place. We shall explain how to check the straightness of a ruler at the end of this chapter, but the problem of drawing a *straight line* will have to wait until chapter 2.

To illustrate further the spirit of trial and experiment in mathematics, a very simple model of this geometry can be made on a wooden board with two pins and a standard set square from a geometry set. Draw a semicircle with AB as diameter, and stick the pins firmly and vertically at A and B. Slide the 90° corner against these pins and observe that it follows the semicircle exactly, just as was expected. Now turn the square and repeat the exercise with the 30°, 60° and 45° corners, all of which can be found on various set squares. What is the curve along which the corner moves?

All the curves look like arcs of circles of different radii, with AB no longer a diameter: they are all larger. Next, make a 'square' from a piece of card or plywood with a corner angle greater than 90°. The resulting arc still looks circular and of much larger radius than any of the others.

Two questions now arise. Are the arcs truly circular and how might we complete them to form the circles, if in fact they are all circular arcs? Figure 1.4 shows the arc produced by a corner of angle 45°. By symmetry, if it is circular the centre must lie on the perpendicular bisector of the line AB, which we might as well take to be the y-axis. If we do indeed have such a circle, then using theorem 1.1, we draw AQ to be perpendicular to AP. Where it crosses the y-axis we mark Q. We push in pins at P and at Q, and sliding a 90° square against these should produce the required circular arc.

Without detailed working, if angle APB is $t°$ and AB = $2l$, the equation of the circle is

$$x^2 + \left(y - \frac{l}{\tan(t)}\right)^2 = \frac{l^2}{\sin^2(t)}.$$

Having drawn this circle we can now justify why the point P, constrained to move by pins at A and B, really does move around it using a generalization of theorem 1.1. This generalization says that if we take a chord of a circle, such as AB, and any point on the circle, such as P of figure 1.1, then the angle APB is constant.

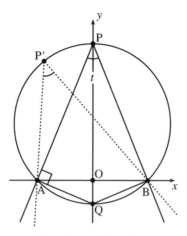

Figure 1.4. Extending theorem 1.1.

In fact, the theorem tells us what that angle is, but that is another story.

Inspection of the figure shows that angle AQB must be 135°, and a 'square' with this corner could be used to draw the part of the circle below the x-axis. In general, if the angle APB is $t°$ we can see that the angle of the other square should be $(180 - t)°$. This is hardly a practical way of drawing a *circle*, but before dismissing it completely two pins and a 'square' of an angle approaching 180°, say 165° for example, could be used to draw short *arcs* of circles with large radii, which is itself an important application. As we shall see later, there is another way of achieving this but at the expense of much more complicated linkages.

1.1 Cutting Lines

It would be a strange book about mathematics that began by telling its readers that all its drawings, such as those of figure 1.1, were at best only approximations to the truth. However this necessarily follows from Euclid's definition of points and lines. Heath translated the opening of Euclid's *Elements*, book I, with the immortal sentences:

> A *point* is that which has no part. A *line* is a breadthless length.

The idea of such a breadthless length is, of course, a mathematical fantasy, for in practice every line must have some width, otherwise it would be invisible! So this makes drawing a line actually a little more tricky than one might suppose.

Had this book been concerned only with pure geometry, these remarks would not have been necessary since the mathematical definitions integral to geometry are understood and accepted. While Hobbes (1588–1679) tried in vain to develop a *theory* in which lines have width, our concern in this chapter is more *practical*. Lines are used for all sorts of other purposes—art, dressmaking patterns and electrical circuit diagrams are just some examples. In the rest of this chapter we discuss the physical and practical consequences of having to work with broad lines.

1.2   The Pythagorean Theorem

A problem which anyone who has done some home improvement is bound to have stumbled across is that of cutting out a shape. For example, when sawing plywood some material is lost in the cut. We shall illustrate this problem and explain one solution with models, the first of which encapsulates the elegant demonstration of the Pythagorean theorem given by Kelland in 1864.

This theorem, also known as Euclid, book I, proposition 47, is perhaps the most famous mathematical result. As everyone learns at school, this relates the lengths of sides in a right-angled planar triangle. In particular, it states that for any right-angled planar triangle with sides of length a, b and c, as shown below,

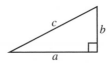

we have the following relationship between the lengths:

$$a^2 + b^2 = c^2.$$

This is an algebraic statement of the theorem. A geometrical statement is that 'the area of the square on the hypotenuse is

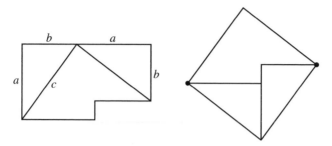

Figure 1.5. The Pythagorean theorem according to Kelland.

equal to the sum of the areas of the squares on the other two sides'. Both amount to the same thing, but it is through this geometrical statement that we obtain a physical demonstration by constructing, dissecting and comparing physical areas.

Kelland's physical model, or jigsaw, consisting of only three pieces, is shown in figure 1.5. On the left we have the two squares, b^2 and a^2, joined. These are made from three pieces, two of which are copies of the original triangle. These can then be rearranged on the right to give a square of area c^2.

Actually, what is particularly satisfying about this dissection is that the pieces may be hinged. One arrangement of the hinges is shown by the dots on the right-hand figure. When hinged in this way the pieces may be moved continuously from one configuration to the other and this makes a most excellent model. Not only is this much more satisfying than a jigsaw, but the pieces are much less likely to become lost. In practice, card with thread attached carefully by sticky tape makes an acceptable, if not a robust, model. Wood or plastic is much more durable, of course, and the classical solution is to use polished hardwood with inlaid brass hinges. However, when cutting wood or plastic, we have to be careful about the material lost by the width of the saw blade and it is to the solution of this problem we now direct our attention.

We start by drawing the left-hand diagram of figure 1.5 on a piece of plywood thick enough to allow hinges to be fixed. Although it is tempting to start with a piece of wood with two straight edges at right angles it is better to work the whole construction on a piece that is larger all round.

Before we cut any lines the centres of the two hinges must be marked and drilled so that their centres will fall exactly at the centres of the intersections of the lines. The next step is to cut out the joined squares. There are two possible approaches we could use. The first is to saw out a slightly oversized piece and sand down to the outer edge of the lines. The second is to saw down the centres of the lines. Since we can only saw down the two diagonals along their centres it is this second method that is needed for all cuts, otherwise there will be noticeable changes in sizes between the parts where they meet.

A saw whose *kerf*, i.e. width of cut, is twice the thickness of veneer, available in handicraft shops, is ideal. Once all the sawn edges are covered with veneer the original shape will be restored, and they will consequently fit together well in both configurations.

It may be thought, with some justification, that the procedures we have outlined above to make the model of this mathematical dissection are over elaborate and too demanding. This is true enough for a simple demonstration model, but to exhibit your craftsmanship to its fullest extent they are worth following. With a little care the lines where the pieces meet will be better approximations to Euclidean lines than those on the original drawing.

This form of construction was followed exactly in the hinged model of Dudeney's dissection, which is a dissection of an equilateral triangle into pieces which can be rearranged into a square. The solution to this problem is often attributed to Henry Ernest Dudeney (1857–1930), one of the greatest nineteenth-century puzzlers. This dissection appears in his book (Dudeney 1907) as the haberdasher's problem, although it had previously been posed in his puzzle column in *Weekly Dispatch* on 6 April 1902. Two weeks later he reported that many people had spotted that this was possible using five pieces, beginning by cutting the triangle into two right-angled triangles. The only person to correctly send in a solution using *four* pieces was a Mr C. W. McElroy of Manchester. Intriguingly, Frederickson (2002) leaves us in doubt as to whether Dudeney knew how to solve the puzzle using only four pieces before he posed it.

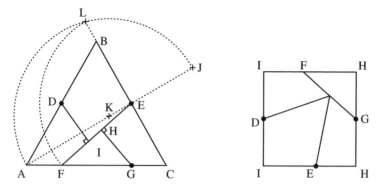

Figure 1.6. Dudeney's dissection.

The dissection of an equilateral triangle into just four pieces, which can be rearranged into a square, is achieved as follows. Begin with an equilateral triangle ABC with sides of length l. A calculation shows that the area of this shape is

$$\frac{\sqrt{3}}{4}l^2. \tag{1.1}$$

Hence, the square will have sides of length $\frac{\sqrt[4]{3}}{2}l$. For our purposes we do not wish to measure these lengths, which would be inaccurate. Rather than taking a metrical approach, we would like to use *geometrical constructions* with only a straight edge and a pair of compasses. If you are unsure how to actually carry out these constructions, then skip ahead to chapter 4 where we examine the topic in detail.

We begin by marking the points D and E midway along the sides AB and BC. Extend the line AE to J so that EJ equals EB. Bisect the line AJ at the point K, and draw an arc centred at K through A and J. Extend EB to L on this arc. Now, with E as centre, draw an arc of radius EL. Where this crosses the line AC mark a point F. In fact EF is the length of the side of the square. Make FG equal to EB. Drop a perpendicular to EF through G, and denote by H where these two lines intersect. Lastly, drop a perpendicular to EF through D, and denote by I where these two lines intersect. Note that the distances FH and EI are equal. This is shown in figure 1.6.

Figure 1.7. Extending lines to complete the corner.

Just as with Kelland's demonstration of the Pythagorean theorem, it is particularly satisfying that the pieces may be *hinged*. By placing hinges at, for example, the points D, E and G, it is possible to transform one shape continuously into the other. This makes it possible to produce a table, a teapot stand or other objects which are useful for two, three, four or six people. An example of a teapot stand is shown in plate 1. What is even more surprising about this particular dissection is that the grain of the wood lines up correctly in *both configurations*.

1.3   Broad Lines

The marking out of sports pitches and tennis courts are two familiar examples of using broad lines. Lines of the width suitable for geometrical drawings would be useless as they would be invisible to player, referee and spectator. In these two examples lines meet at right angles, and so we must consider what happens at their junction. The probable result is illustrated in exaggerated form on the left of figure 1.7. Notice the annoying missing part, caused by the intersection of the rectangles which represent the lines. What we really would prefer here, but by no means always, is for the lines to extend a little beyond where they should, to give a corner as shown on the right of the figure.

A consequence of having broad lines is the need for very careful interpretation of what is meant by 'over the line'. Traffic police and soccer referees clearly have different views about its meaning when considering whether or not to prosecute a motorist or to allow a goal. Depending on the interpretation, the broad lines might be redrawn to make explicit a choice from the two possibilities shown in figure 1.8, which illustrate the fact that when two lines meet, the coordinate should really label the corner of the shape and not the middle of the line.

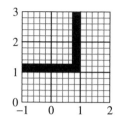 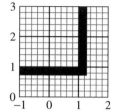

Figure 1.8. Emphasizing the inside and outside of a corner.

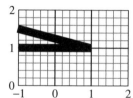 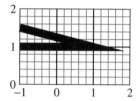

Figure 1.9. A more acute problem.

Of course, not all lines are perpendicular, as is the case with some standard road markings, and in particular the isosceles triangle found at road junctions. Here the problem of the lines meeting at a corner is even more acute (excuse the pun). Two options for drawing this are shown in figure 1.9. On the left are two lines drawn independent of each other and on the right is the shape obtained if we extend the lines to fill in the gap as we did in figure 1.7. While extending the lines appeared to be a sensible policy for the right angle, a small angle between the lines leads to a huge overshoot here.

The problem of cutting a 'vee' notch is exactly the same as that already noted with road markings. If the marked angle is 90° or greater, two straight saw cuts with the blade perpendicular to the wood are adequate. If the angle is less than 90° it is impossible to saw right into the notch of the 'vee', and a different tool, such as a chisel, will be necessary to cut out the final piece.

A completely different approach is demonstrated by two standard pieces of mechanical workshop equipment: a vee-block and a centre drill. Blocks are used to support work in milling machines and on drill tables, and usually come in matching pairs with ground faces. One block is shown in figure 10.18.

To overcome the problem of machining right into the hollow of the vee it is usual practice to cut a slot, as is shown below. In this way square work, for example, can be held firmly:

The same problem arises with the conical centre used to support the end of a long slender piece of work at the end of a lathe. It is easy enough to turn the conical centre itself, but it is extremely difficult to drill the matching hole in the workpiece. As with vee-blocks the solution is not to attempt to drill to a point, but to remove the point altogether with a special centre drill so that the cone can sit properly, without shaking.

A true cone can only be turned on the end of a round bar if the tool is at centre height. If it is low then the result is not a cone, not even a truncated cone, but a section of a hyperbola of revolution: a hyperboloid. We leave this as an exercise for you to prove.

1.4 Cutting Lines

Although it is very unlikely to be found in a home workshop, we discuss some of the geometrical problems that arise when using a computerized numerical control (CNC) milling machine. The machine tool is essentially a vertical milling machine with the cutting tool rotating about a vertical axis whose height, the z-coordinate, is controlled together with the x- and y-coordinates of the work table.

For simplicity we assume that the height of the tool is fixed and we wish to cut to some defined curve on a piece of metal plate to make a template. Examples of this can be seen in figure 10.5. The milling cutter is similar to a drill but much more rigid and uses both its end and side faces to cut away material. When moved sideways it is effectively drawing a broad line.

Figure 1.10. Cutting a circle and a straight line.

It turns out that there are only two curves for which the tool's path has the same *form* as the desired curve: they are the circle and the (closely related) straight line. The reason for this is that the evolute, which is the locus of the centres of curvature, of a circle is a point at its centre. For a straight line the equivalent point lies an infinite distance away. For a circle, then, the tool's path is another circle with the same centre and whose radius is increased by the radius of the cutting tool. A straight line obviously has a straight-line tool path, displaced again by the radius of the cutting tool, as illustrated in figure 1.10. No other curves have a point evolute, and so tool paths are necessarily of a different shape from the desired form.

Returning briefly to the drawing board, you may well be familiar with sets of radius curves used for drawing arcs of circles of small radius. Careful measurement of the sizes show that each is smaller by $\frac{1}{32}$ in than its stated size to allow for broad pencils or pens.

After a circle and a straight line one would argue that the mathematically simplest shapes are the conic sections. These include the ellipse and the parabola, and we shall meet all these shapes again—for example, in chapter 2 we show how two identical parabolic templates can be used to draw a straight line. Here we concentrate on cutting out a parabolic shape. As one example, let us see what curve a tool of radius R should take to produce an artefact where the boundary has the equation

$$x^2 = 4ay \text{ mm}^2.$$

Because we are about to make a real template the equation of the parabola above must have its dimensions clearly defined, otherwise it is meaningless.

At some point $P = P(x, y)$ on the parabola the centre of the tool is at $T = T(X, Y)$. The problem is to relate the two sets

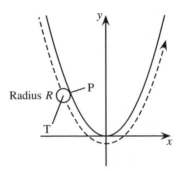

Figure 1.11. Cutting out a parabola on a CNC machine.

of coordinates. This relation can be described in language by saying that both T and P are moving in the same direction, a fixed distance apart. This is expressed in mathematical terms by equating the tangents of the curves. Second, we know that P and T are a distance R apart, for which we return to our friend the Pythagorean theorem. Lastly, since the tool and the work surface are tangential we have the normal line to the parabola passing through both (x, y) and (X, Y). No explicit attempt has been made here to solve the resulting equations to find the fundamental relation between X and Y, except to note that it is not simple. We have illustrated the solution in figure 1.11. Indeed, although in this case it is possible to solve the resulting system of differential equations exactly this is generally not the case. Hence in practice X and Y would be approximated with straight-line segments for some given, short, step length. Modern and necessarily expensive machines have the inbuilt facility to do this provided that they are given the equations of the curve and the tool radius. Most CNC machine centres have the capacity to cut straight or circular arcs after being given the tool radius.

This discussion of CNC machining has been included not because such tools are likely to be in a home workshop, but to show how apparently simple tasks such as cutting a shape like a parabola can lead to real geometric problems. The same can be said of a CNC lathe: the solids of constant breadth we examine in chapter 10 were turned this way. All the arcs are circular, but had the arcs not been of this form, the programming would have been much more difficult and tedious.

We need to look at this geometry in a reverse sense when considering what shape of cam and its follower should be used to provide some specified linear movement. The follower is effectively a point bearing on the face of the cam. In reality the follower is not quite a point, it is more typically in the form of a circular arc. In effect the tool path is dictated by the motion required of the follower and this in turn defines the shape of the cam.

1.5   Trial by Trials

This chapter has really been about the interpretation of lines and marking out work prior to cutting. It is very appropriate then that it finishes by describing how three of the essential pieces of workshop equipment can be made. Although chapter 2 shows how straight lines can be drawn by link-work they are hardly practicable when marking out in the workshop. What is needed is a straight edge. These can be obtained commercially, and for home use, at least, steel rules are not only straight but also graduated, something we address in chapter 4. In order to make one straight edge from scratch with no reference edges or planes available is effectively impossible. However, if *three* are made together they can be checked against each other and their ultimate straightness ensured.

Consider a first attempt A, shown in a misaligned form in figure 1.12. It is not straight so make another B. A and B fit together perfectly when they slide over each other. The only conclusion is that either they are both straight or they are both arcs of some circle of large radius. Their straightness is only confirmed if a third edge is made, C, and all three pairings of edges in all positions match. Then, and only then, are they confirmed to be straight.

A similar argument is applied to making a truly flat surface, usually called a surface plate if designed for use on a bench or a surface table for larger stand-alone models.

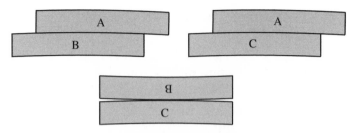

Figure 1.12. Comparing three edges.

Lastly we consider an engineer's 90° square. Again, three are made, A, B and C. When the stocks are supported on a straight edge or surface plate and match over the full length of the blades we therefore have for the corresponding angles

$$\alpha + \beta = 180°,$$
$$\beta + \gamma = 180°,$$
$$\gamma + \alpha = 180°,$$

so that $\alpha = \beta = \gamma = 90°$.

Squares, like the edges and surface plates, are manufactured to very high degrees of precision. British Standard 939 ('Engineers' squares (including cylindrical and block squares)', 1977, British Standards Institute) specifies the tolerances for squares and these are given in the seven tables within the specification. What is perhaps curious at a first reading is that it is only after the first five tolerances have been satisfied that it is possible to define the tolerances of the 90° angle. Even this is not specified as an angle, but as a linear departure from squareness between the stock and the blade.

Chapter 2

HOW TO DRAW A STRAIGHT LINE

The straight line is a stronger and more profound expression than the curve.

Piet Mondrian

This chapter heading comes from a book by A. B. Kempe, *How to Draw a Straight Line: A Lecture on Linkages* (Kempe 1877). The charm and gentle tone of the book suggest that the original lectures must have been a delightful experience. If you ask most people what a straight line is their answer will probably involve 'the shortest distance between two points'. Of course they do not always mean this since the shortest distance to the Antipodes from Yorkshire is downwards through the centre of the Earth: these two places are at opposite ends of a diameter. What they probably mean is the arc of a path on the surface of the Earth which is a great circle. In this chapter we shall adopt the usual geometers' convention and claim that the Earth is flat. By restricting ourselves in this way, we return to the usual idea of a straight line in a *plane.* The obvious way to draw such a straight line is to use a rule or straight edge but, of course, this argument is circular and Kempe (1877) makes much play of the need to make the 'first' straight line. More serious practical problems arise if the rule is simply not long enough or if it is needed not to guide a pen or pencil but a piece of heavy moving machinery. Here we focus the discussion on the application of linkages for this latter purpose although, to be true to the chapter heading, we include a section on how to make models of linkages that can be used to draw a straight line using a variety of techniques.

To appreciate fully the significance of linkages it is necessary to go back over 200 years to examine the development of steam

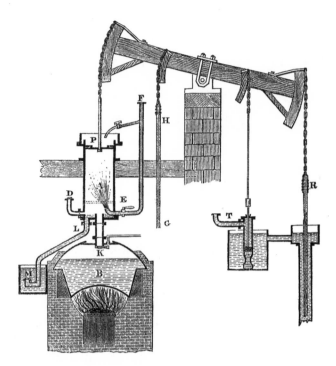

Figure 2.1. A chain used in beam engine transmission.

power and, in particular, the beam engine, to see why these linkages played an essential part in the Industrial Revolution that led
to the economic growth and prosperity of the United Kingdom
in the eighteenth and nineteenth centuries.

Figure 2.1 shows a very much simplified beam engine. The
vertical cylinder, with its piston P and rod, transmits power to
rock a beam at the other end of which there was commonly a
long rod R driving a pump, typically to clear water from a mine.
In very early beam engines, such as that shown, the steam in the
cylinder condensed as the valve at E was opened to let in cold
water. This created a vacuum in the cylinder and it was the force
of atmospheric pressure that pushed the piston down. It was
only later that the direct expansion of steam created the force
which drove the piston.

In any case, the piston rod is designed only to take axial
loads, that is to say a push or pull. Its generally slender form
is not designed to resist sideways loading forces and so it must

somehow be constrained to move in a straight line. However, the end of the beam to which it must be connected can only move in the arc of a circle. If the piston rod were connected directly to the beam then the piston rod would be subjected to sideways forces and the engine would rapidly wear out, assuming of course that it even worked in the first place.

To experience, literally first hand, some of the consequences of not sending the piston rod in a straight line take a cafetière and, to be on the safe side, fill it with cold water rather than hot coffee. If you can manage to depress the plunger carefully so as to keep contact between the rod and the hole in the lid to a minimum you have achieved the ideal: the rod has not been subjected to any lateral forces and the wear on it and the lid of the cafetière is at a minimum. Repeat the exercise, but this time do it more hurriedly: the rod will rub against the lid and often cause it to tip up or even come off the top completely. In extreme cases the whole body of the cafetière might tip and spill some of the contents. Now change some of the words in this paragraph; substituting 'stuffing box' for 'hole in the lid', 'cylinder cover' for 'lid', and 'cylinder' for 'body of the cafetière' and you have reproduced exactly what could happen in a beam engine. You will have demonstrated the need for some form of guidance for the piston rod. You will also have noticed that the diameter of the hole in the lid is slightly larger than the plunger rod. Aside from the fact that cafetières are mass-produced non-precision goods, this clearance, or play, illustrates an important aspect of design in all pivots: there must be some clearance, otherwise all joints would be too stiff to move and lubrication would be impossible. The significance of this apparently trite and obvious observation is that in any system of links there must inevitably be some play or tolerance in the movement of the pivot designed to provide the straight-line guide. This implies that the geometry of a linkage need not describe an *exact* straight line provided its theoretical departure from true linearity lies within the practical tolerances or play of the system to which it is being applied.

Incidentally, the need for straight-line linkages did not arise with beam engine design because the first engines were single

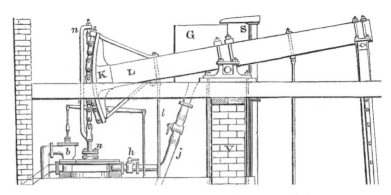

Figure 2.2. Francis Thompson's opposed chains.

acting: that is, power was developed only on the downstroke of the piston and the return upstroke was achieved by simply balancing the beam and pump to pull the piston up. When the piston was moving down it was pulling the beam, on the upstroke the roles were reversed and the beam pulled the piston. In this way the connection between the piston rod and the beam was always under tension and it was quite acceptable to use a chain; the end of the beam was in the form of a large circular arc, i.e. a sector (see figure 2.1). You can still see this mode of working in a 'nodding donkey', used for pumping oil from wells.

One of James Watt's many innovations and improvements to engine design was to have double-acting engines. Power was then developed on both the upstroke and the downstroke and a chain connection between rod and beam was clearly no longer a possibility—you can pull a chain but you cannot push with it. One solution was to use two chains, and this option was taken by Francis Thompson in Ashover in around 1783. His atmospheric engine, part of which is shown in figure 2.2, had two separate and opposed cylinders. The two chains worked under tension in opposite directions.

Another solution was to fit teeth to the piston rod engaging with pegs on the beam. Neither of these solutions was entirely satisfactory, the additional chain increased the height of the engine and the rack was noisy. Workshop technology at that time, the 1780s, was not capable of producing long straight guides and although they could have been made in sections, their

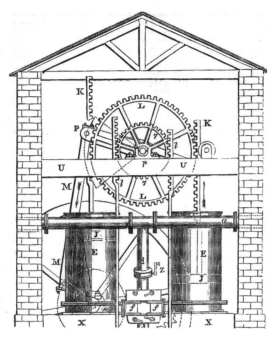

Figure 2.3. Teeth on the piston engaging with pegs on the beam.

cost of manufacture and the cost of erection would have been so high that beam engines would have been totally uneconomic.

This is a very brief introduction to the problem of producing straight-line guides to take advantage of the effectively doubled power output of double-acting steam engines. Although beam engines have long been superseded, the need for such linkages still exists in a variety of engineering applications, and linkages for other straight-line motions are part of everyday living. Look at the hinges on cars, modern windows and windscreen wipers. Rarely do they use simple hinges.

The original name for these was 'parallel motion' linkages for reasons that are not exactly clear, although this term was widely used as you will see if you ever look in an old book about steam power or mechanical engineering. No attempt is made here to try to describe the whole range of linkages that have been invented over the years, but rather we present a selection of the best known, practically important or geometrically most interesting. We have divided the linkages into three categories.

The first category is of *approximate* straight lines. Although the idea of an approximate straight line sits uneasily, as mentioned above it is often a perfectly acceptable solution to a real engineering problem. It should be emphasized that departures from a true straight line are inherent in the geometry of the linkage and not a consequence of either poor workmanship or deliberate play in the joints. In the second category of linkages the geometry is such that if they are correctly made they will generate an *exact* straight line.

The last category comprises *guide linkages*. These are hybrid linkages in the sense that they rely on one of their pivots being itself guided in a straight line. This may almost sound like cheating, but their importance is that they magnify this guided movement so that a part of one link moves in a straight line but over a much greater distance. It is much cheaper to make a short straight guide and use a linkage to increase its range than it is to machine a long guide.

2.1 Approximate-Straight-Line Linkages

James Watt (1736–1819) is remembered today for his pioneering work on steam engines, and his name is commemorated in the unit of power. His own view of his life's work does not quite agree with current opinion: in old age he wrote to Matthew Boulton that

> [a]lthough I am not over anxious after fame, yet I am more proud of the parallel motion than of any other invention I have ever made.

It was James Watt who invented the first straight-line linkage, and the idea of its genesis using links is contained in a letter he wrote to Boulton in June 1784.

> I have got a glimpse of a method of causing a piston rod to move up and down perpendicularly by only fixing it to a piece of iron upon the beam, without chains or perpendicular guides... and one of the most ingenious simple pieces of mechanics I have invented.

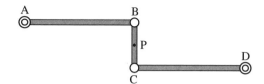

Figure 2.4. A schematic of Watt's linkage.

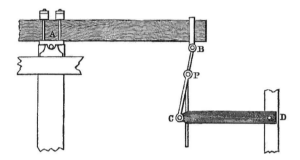

Figure 2.5. The general form of Watt's linkage.

Watt published his linkage in a patent dated 24 August 1784, and it is important to remember that he did not claim it produced a true straight line. In a letter to Boulton on 11 September 1784 he describes the linkage as follows.

> The convexities of the arches, lying in contrary directions, there is a certain point in the connecting-lever, which has very little sensible variation from a straight line.

In its simplest form Watt's linkage consists of just the following three bars, AB, BC and CD (see figure 2.4). Here A and D are fixed in space, but free to rotate, forcing B and C to move in arcs of circles, so that the two long arms AB and CD form the 'arches' to which Watt refers. Links AB and CD are the same length, and the centre of the short connecting link BC, labelled P, is the part that moves vertically in the approximate-straight-line manner. It is worth noting that the length of AB does not have any particular relation to the length BC. Watt's linkage also has a more general form with AB ≠ CD, illustrated in figure 2.5,

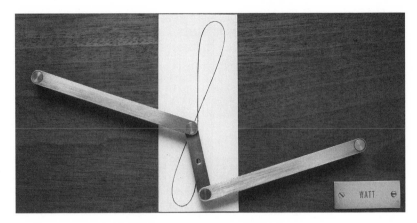

Figure 2.6. Watt's linkage.

in which P is positioned with the ratio

$$\frac{AB}{CD} = \frac{CP}{CB}.$$

The longer the arms AB and CD are to BC, the better the path of P will approximate a straight line. Note that this result is only approximate, and only for small displacements of B, or equivalently C. The complete path taken by the point P is shown in figure 2.6.

Before going any further with Watt's linkage we ought to look at the use of this linkage by Phineas Crowther (1763–1818). On paper it appears to be the same, and instead of drawing another figure, figure 2.6 showing Watt's linkage is adequate. Crowther's objective was to dispense with the beam by placing the crankshaft directly above the cylinder, the piston being constrained with a Watt-like linkage. The only real difference then is one of application: Watt's was for beam engines whereas Crowther's was constructed for conventional engines. He was an engineer from the northeast of England and the main application of his linkages was to winding engines, such as that shown in figure 2.7, made for hauling coal or ore wagons up inclines. We mention this because one of his working engines has been preserved and re-erected in the National Railway Museum in York and is routinely shown working, albeit with an electric drive, at frequent intervals throughout the day. According to Watkins

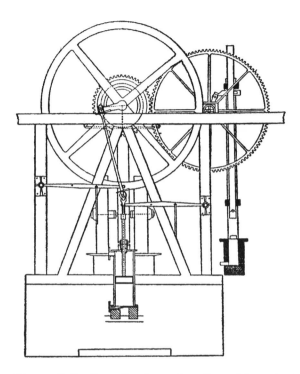

Figure 2.7. Crowther's engine, from his patent
(UK patent number 2378) of 1800.

(1953), such engines were popular and some were in continual
use for decades.

The dimensions of this engine are known, so we use it as an
example to show the movement of the piston rod, and in particular its departure from movement in a straight line. Since it is an
old engine, the original imperial measurements have been used
in the calculation. There is one slight difference from the Watt-
type linkage shown in figure 2.4. At the centre of the stroke,
when the arms AB and CD are horizontal, the centre link BC
is not vertical but approximately $\theta = 5°$ from the vertical. The
dimensions are as follows:

$$AB = CD = 132 \text{ in},$$
$$BC = 27 \text{ in, with P at the centre},$$
$$\text{stroke} = 60 \text{ in}.$$

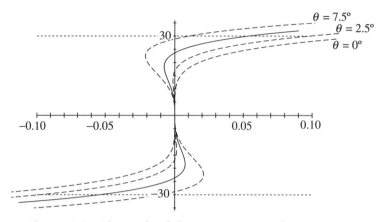

Figure 2.8. The path of the pivot in Crowther's engine.

The horizontal distance between the fixed pivots at A and D is

$$2AB - BC\sin(5°) \approx 237.7 \text{ in,}$$

and the vertical separation is

$$BC\cos(5°) \approx 26.9 \text{ in.}$$

We can calculate the path taken by the point P as it travels through a complete stroke of 60 in, and this is shown in figure 2.8 with a solid line. We have repeated these calculations for values of $\theta \neq 5°$ and included these using dashed lines for the purposes of comparison. Notice the vastly exaggerated horizontal scale of ± 0.1 units, compared with the vertical scale of ± 30 units. Hence, the path of P is actually quite a good approximation to a straight line, despite the appearance of figure 2.8.

What is clear is that the linkage is very sensitive to the positioning of the fixed pivots, and if the original 5° departure from vertical were to be rotated slightly, the departure of the path of P from a straight line would be greater at the ends of the stroke. Even so, the departure of ± 0.05 in at the ends of the stroke of 60 in, or 0.083%, is quite acceptable given the number of pivots and the amount of play in each.

The Crowther linkage is quite obvious on the engine, but on a beam engine it is sometimes difficult to distinguish the Watt

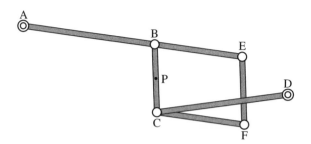

Figure 2.9. Watt's complete parallel motion.

linkage from the other parts. Furthermore, the beam is quite long and an arm of equal length would take up a prohibitive amount of space. To overcome this problem Watt developed the basic linkage into his 'complete parallel motion', also included in his original 1784 patent. While the original simple motion is rarely seen on beam engines, the complete parallel motion is much more common.

In the arrangement shown in figure 2.9 we have the main beam of the engine with its central pivot at A just as before. However, the point B is not at the end of the beam, which is E, but part way along. The rod CD, which is equal in length to AB and is known as the *radius rod*, is connected to the beam by a shorter link BC. The centre of this, P, will move in an approximate straight line. Here the link ABCD is exactly Watt's or Crowther's linkage as we had it before. The other links form a parallelogram BCEF, and the pivot F is connected to the piston rod. You are encouraged to consider the motion of F, which is of course critical. The most straightforward case is when the length AB equals that of BE, which is not the same as that illustrated in figure 2.9.

This figure also suggests why the term 'parallel motion' was used originally, because apart from having this more compact form, the two uprights in the parallelogram guide both the piston rod and the valve gears. An example of this is shown in figure 2.10. Here an extra vertical rod is attached to the driving rods of an air pump. Air in steam is a nuisance as it is not condensed and can therefore lead to a loss of power. Dickinson (1963) is a good starting point for more information about this and other practical aspects of steam engine history.

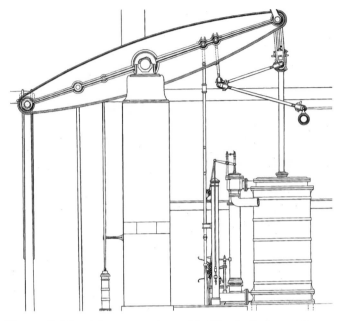

Figure 2.10. A beam engine with Watt's linkage shown.

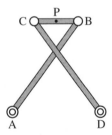

Figure 2.11. Chebyshev's linkage.

Following Watt's discovery a whole range of linkages was developed, and here we must limit the discussion to some of the more interesting. The first we consider was developed by the Russian mathematician Pafnuty Chebyshev (1821–94), who was fascinated by linkages. For many years he was of the firm belief that linkages could never be designed to produce exact straight lines. His best-known linkage is shown in figure 2.11.

In essence this consists of the three bars of Watt's linkage with the radius arms crossed, but it is quite different in the strict requirement that the lengths of the links must be in the following proportions:

$$AD = 4,$$
$$AB = CD = 5,$$
$$CB = 2, \quad \text{with P at its centre.}$$

In the mid position, as drawn, the linkage is symmetrical and the Pythagorean theorem can be used to show that P is 4 units above the baseline AD. When either of the linkages AB or CD is in the vertical position, P is again exactly 4 units above AB. It is well worth making a model of this linkage and looking at the complete path of P since there is an interesting curve between these positions. The complete curve traced is shown on the right-hand side of plate 2, although you will immediately notice we have used a different linkage here. More on this in a moment.

For now notice that the curve appears to be made up of two parts, one of which appears to be straight, the other curved. The path of P does not actually follow a straight line, and figure 2.12 shows this part of the movement. Please note carefully that the vertical axis has been exaggerated wildly. The maximum deviation from the horizontal straight line $y = 4$ is less than 0.01 vertical units in a horizontal movement of 4 units, or 0.25%. Now, to put this departure from a true straight line in context, if we take the distance AD to be 100 mm, then the maximum departure from a straight line is 0.25 mm. Recall that we are forced to use broad lines in any sketch, and a common printed line width is around 0.2 mm. If you have a propelling pencil, this is likely to use graphite of 0.5 mm. If we draw the x- and y-axes on the same scale then the maximum theoretical deviation is about the same as the broad line used to draw it, so it will be hard to distinguish. We have tried to show this accurately in figure 2.12. The top is exaggerated and the bottom uses the same horizontal and vertical scales. Whether you can perceive a difference between the bottom two lines will depend to a large extent on the quality of the printing used to create this page!

There appears to be no connection between Chebyshev's linkage and the arrangement of links shown in plate 2. However,

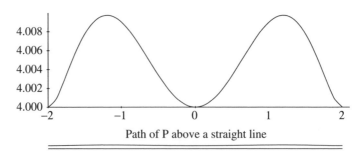

Figure 2.12. Chebyshev's approximate straight line.

we shall now show that the two arrangements produce identical curves, and we will show that the path of the second form can be calculated more easily and directly.

First draw Chebyshev's linkage ABCD as before, as shown in the left-hand diagram of figure 2.13. Mark P as the midpoint of BC. Next draw a similar linkage A'B'C'D' of half the size, with the same orientation, with D = D', and with A' on the line AD. Hence CP is parallel to B'C' and these links are equal in length (one unit). Now draw a line through PB' and extend this so the length equals CD. Then we have a parallelogram PCDE. In particular, as B and C move on a circle, with centre at P, E will move on a circle with a centre at D.

Conversely, if E moves on a circle of radius 1 with centre at D, then the three-bar linkage A'B', PE, ED forces P to move along the midpoint of the line BC, where B and C are part of Chebyshev's original linkage. Hence, if we construct an arrangement consisting only of DE, A'B' and EP, we know that P will follow the path of Chebyshev's original linkage. This is the alternative form shown in plate 2.

In the original form of the linkage the point P crosses over the two links AB and CD during one complete motion. In the second form the point P is unobstructed, and the motion of the linkage can be driven by a circular motion of E. Alternatively, the straightest part of the curve can be obtained by reciprocating the motion of E on a circular arc. Both of these aspects are significant practical considerations, making the second form much more useful. It is also easier to calculate the path of P from geometrical

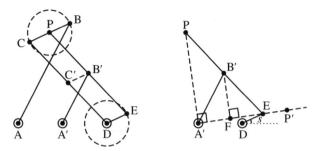

Figure 2.13. Chebyshev's original and alternative linkages superimposed.

considerations in this second form, since we have that A′B′ = B′P = B′E.

We begin our detailed examination of Chebyshev's alternative linkage by providing a new diagram, without the original linkage in place. This is shown on the right-hand side of figure 2.13.

We recall that A′B′ = B′P = B′E. Furthermore, PA′E is a right-angled triangle with right angle at A′. If we drop a perpendicular from B′ onto A′E, and mark this point as F, then, by similar triangles,

$$\frac{A'P}{A'E} = \frac{B'F}{FE} = 2\frac{B'F}{A'E}.$$

So that

$$A'P = 2B'F = 2\sqrt{(A'B')^2 - \tfrac{1}{4}(A'E)^2} = \sqrt{4(A'B')^2 - (A'E)^2}.$$

Extending the line A′E if necessary, mark P′ on A′E so that A′P = A′P′. We take the origin to be A′ and E to have coordinates (x, y), then P′ has coordinates

$$P' = \left(\frac{A'P}{A'E}x, \frac{A'P}{A'E}y\right),$$

and by a direct transformation, P will have coordinates

$$P = \left(-\frac{A'P}{A'E}y, \frac{A'P}{A'E}x\right).$$

By taking

$$E = (x, y) = (A'D + DE\cos(s), DE\sin(s)),$$

 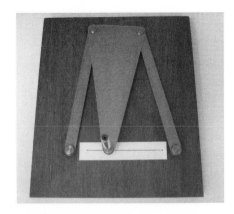

Figure 2.14. Roberts's linkage.

and noting that $(A'E)^2 = x^2 + y^2$, we can put all the pieces together to obtain an explicit formula for the path P in terms only of the lengths of the links, and the angle s. As you can see, this does not result in a satisfying, simple formula. It is a great pity that mechanisms as simple as these linkages, in either original or more general forms, do not yield neat mathematical descriptions in elementary terms. Surprisingly there are many other arrangements of linkages which will also produce exactly this curve.

A further development was made by Richard Roberts (1789–1864). It is another example of a Watt-type linkage, and here the restrictions on link lengths can be relaxed. All that is necessary is that

$$AB = BP = CP = CD$$

and

$$BC = \tfrac{1}{2}AD.$$

A model of Chebyshev's crossed linkage can be modified by uncrossing the links and then adding an arm. An example is shown in figure 2.14, with an accompanying schematic.

The trace of P as it moves between A and D is a very close approximation to a straight line. For the purposes of comparison with Chebyshev's straight line in figure 2.12 we take exactly the

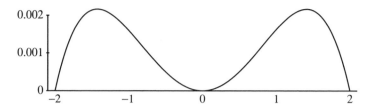

Figure 2.15. Roberts's approximate straight line.

same linkages, supplemented with the extra bar in figure 2.14. Then, without details of the calculation, the path of P is shown in figure 2.15, showing a substantial improvement. Again, the vertical axis has been exaggerated wildly and the maximum deviation from the horizontal straight line $y = 0$ is less than 0.0022 vertical units in a horizontal movement of 4 units, or 0.06%.

All these examples of Watt-type linkages involve two fixed points and three bars. In some sense the two fixed points lie on a fourth bar. Hence we refer to these as four-bar linkages. The free ends of the two fixed bars must move in circles, and so we consider a point P relative to the third bar. Notice the progression from a point P on the link between B and C, then on an extension arm in line with BC, and then in Roberts's mechanism to an arbitrary point not on the line BC but in a position fixed relative to it. This is the most general situation possible without the addition of extra linkages.

2.2 Exact-Straight-Line Linkages

Historically, the first of these linkages was described by Pierre Frédéric Sarrus (1798–1861) in 1853. It differs from all the other linkages so far described in this chapter in that its parts move in three dimensions rather than just the two dimensions of planar linkages. Although it may not have been used in steam engines it is applied widely in jacks, elevating platforms and similar devices. The crude model in plate 3 shows that the pivots are more akin to hinges, making it look like a short section of a concertina. Although in its completed form it looks complex, the opened-out photograph shown illustrates how simple the

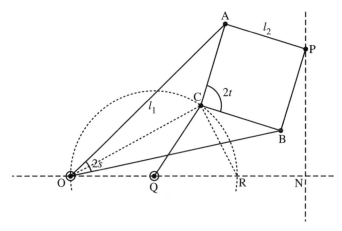

Figure 2.16. Peaucellier's linkage.

mechanism is. A full description of Sarrus's mechanism and other similar hinged systems is given in Dunkerley (1912).

The first *planar* linkage was invented by Charles Nicolas Peaucellier (1832–1913) in 1864 while he was a serving officer in the French army. Two forms of this linkage are shown in plate 4. It uses seven links, and is constructed in figure 2.16.

Without the link CQ we have an arrangement that has become known as the *Peaucellier cell*. The links are such that

$$OA = OB = l_1,$$
$$AP = BP = AC = BC = l_2.$$

For practical convenience, $AC \approx \frac{1}{3}OA$, which determines the maximum opening of the long arms:

$$\sin(s) = \frac{l_2}{l_1}.$$

Before we consider the entire linkage we shall consider carefully this cell, which itself can be broken down into the 'kite' OAPB and the 'dart' OACB shown in figure 2.17.

Using the Pythagorean theorem yet again we have that

$$(OM)^2 + (AM)^2 = l_1^2, \qquad (PM)^2 + (AM)^2 = l_2^2.$$

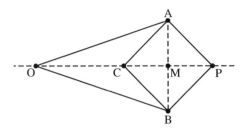

Figure 2.17. The Peaucellier cell.

Subtracting these gives

$$(OM)^2 - (PM)^2 = l_1^2 - l_2^2,$$

and rewriting the left-hand side as a product results in

$$(OM - PM)(OM + PM) = k^2,$$

where k is a constant, and we have used $k^2 = l_1^2 - l_2^2$ for dimensional consistency. However, $(OM - PM)(OM + PM) = (OC)(OP)$ and so we have

$$OC \cdot OP = k^2. \tag{2.1}$$

Hence we have shown that the product of the distances OC and OP is constant. So, if we choose OC, then $OP = k^2/OC$. For this reason, if we have a curve along which OC moves, the curve which P follows is sometimes called an *inverse* of the original. We shall see in a moment that the inverse of a circle can be a straight line, thus solving our problem.

Now, to complete the linkage shown in figure 2.16 we add the final link QC with the condition that $OQ = QC$. We also ensure that O and Q are fixed points, so that the point C moves on a circle, centred at Q. Then, by theorem 1.1 we have that the triangle OCR is a right-angled triangle. Drop a perpendicular from P to the line OQ, and mark the meeting point N, then the triangles OCR and ONP are similar, so that

$$\frac{ON}{OP} = \frac{OC}{OR},$$

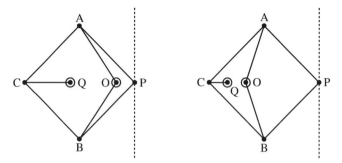

Figure 2.18. Alternative forms of Peaucellier's linkage.

and so

$$\text{ON} = \frac{\text{OC} \cdot \text{OP}}{\text{OR}} = \frac{k^2}{2\text{OQ}} = \text{const.}$$

Since the distance ON is constant, the motion of P must be a straight line, perpendicular to the line OQ.

In a more general case we must retain OQ = QC, but QC does not necessarily have to equal any of the short links l_2. It is a great convenience though in manufacture if the number of different lengths can be kept to a minimum. Peaucellier's linkage is usually shown in the form of figure 2.16, but it can be made more compact by folding it in on itself. Using the same lettering we have the configurations shown in figure 2.18.

The geometry is unchanged, and according to Sylvester (1875a) it was in this folded form that the linkage was used to guide the piston rods on the ventilating engine in the old Houses of Parliament and by a wealthy resident of Southampton for the same purpose on the engine of a steam yacht.

The two key facts about Peaucellier's cell are (i) that O, C and P remain in a straight line and (ii) that the product of the distances OC and OP is constant. This is sometimes called an inversor linkage. Any arrangement with these two characteristics will be sufficient to generate straight-line motion, for we can easily constrain C to move on a circle as before.

While Peaucellier's linkage uses seven links, this number was reduced to just five by H. Hart in 1875, a model of which is shown in plate 5. Hart's linkage uses a cell with just four bars and is

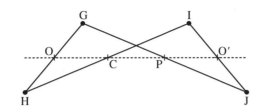

Figure 2.19. Hart's inversor cell.

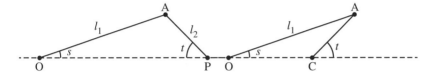

Figure 2.20. Hart's 'kite' and 'dart'.

shown in figure 2.19. This consists of a crossed parallelogram GHIJ, with GJ = HI and GH = IJ. This condition ensures that the (imaginary) lines GI and HJ remain parallel during the movement. Accordingly we shall take any other line parallel to HJ, and on each link we mark a point. We shall show in a moment that the marks on Hart's linkage behave in such a way that the two important properties of Peaucellier's cell are preserved. The diagram shows the case GH/GJ = $\frac{1}{2}$. By similar triangles we have that

$$\frac{GO}{GH} = \frac{GP}{GJ} = \frac{IC}{IH} = \frac{IO'}{IJ}.$$

If the linkage is moved then this property is preserved. Hence OP and CO′ will remain parallel with HJ. Since the distance of OP from HJ is GO/GH, and since this is also the distance of CO′ from HJ, the four points OCPO′ remain collinear during the motion of the linkage. This is the first important property of Peaucellier's cell.

We are left to justify that OC · OP remains constant. One way to do this is to pull apart Peaucellier's cell still further into two components (see figure 2.20). On the left of this figure we have half of a 'kite', and on the right half of a 'dart'. It is then possible to match the dart with OHC and the kite with OGP, and we are done.

A direct argument can also be constructed from a result in geometry by noticing that the four points G, H, I and J of Hart's cell all lie on a circle. In general three points determine a unique circle through them (or perhaps a straight line), so the fourth point being on this circle cannot be taken for granted. Hence, take the unique circle through HGI, then the perpendicular bisector of the chord GI passes through the centre of the circle, and by symmetry the point J also lies on the circle. A result known as Ptolemy's theorem tells us, in the notation adopted for Hart's cell, that

$$GJ \cdot HI = GI \cdot HJ + GH \cdot IJ.$$

Since our links occur in two pairs we have that

$$GJ^2 = GI \cdot HJ + GH^2,$$

or

$$GI \cdot HJ = GJ^2 - GH^2 = k^2$$

for some constant k. By similar triangles, OC is proportional to GI and OP is proportional to HJ, so $OC \cdot OP$ is also proportional to $GI \cdot HJ$, and hence constant. These and other linkages have been given a case-by-case treatment. A general unified approach, in which these two occur as special cases, is given by Kempe (1875).

2.3   Hart's Exact-Straight-Line Mechanism

We include one more mechanism invented by Hart which guarantees exact straight-line motion, and which moves in a most delightful way. It also uses only five bars, in two different lengths. Take five linkages with $AC = BD = a$, $PC = PD = RS = b$ and attach the link RS to R and S so that the distance $RC = DS = c$, where $b^2 = ac$. The sketch in figure 2.21 shows the case where $a = 8$, $b = 4$ and so $c = 2$, which is a particularly convenient selection from a practical point of view. A model of this linkage is shown in plate 6.

Let us ignore the link RS for a moment. If we can arrange the other four links so that angle PCA equals PDB, then the two

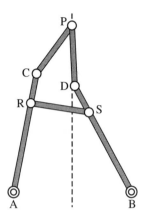

Figure 2.21. Hart's straight-line mechanism.

triangles APC and BPD will be congruent. Hence, AP equals PB and so point P will be found to move along the perpendicular bisector of AB. If we take RS = b and RC = DS = c, then we leave it to you to prove that the angle PCA equals PDB.

Note that Peaucellier's linkage can certainly be used to draw the 'first' straight line, since each link has only two pivots. However, to make either of the linkages shown in plates 5 and 6 it is necessary to create links with three collinear pivots. This presupposes the ability to place three points in a straight line.

2.4   Guide Linkages

This form of linkage relies on one of the pivots being itself guided in a straight line and the function of the other members is to magnify this movement so that one point on one of the links describes a much longer straight line. The point of this system is that although it requires a straight guide, this guide is considerably shorter and therefore easier and cheaper to make than a full-length guide. The best-known example is due to John Scott Russell (1808–82). He was an engineer and naval architect who, among many other activities, was associated with Brunel's large steam ship the Great Eastern, a story told in more detail in Emmerson (1977). Scott Russell's linkage is very closely related to the alternative form of Chebyshev's linkage from section 2.1 and is shown in figure 2.22.

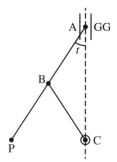

Figure 2.22. Scott Russell's linkage.

Point A slides in a short straight guide GG, fixed point C lies on its centre line, and AB = BP = BC. Initially P is also on the line AC, and then moves so that AP is at some angle t to AC. Then P will have moved $AP \sin(t)$, while A will have moved only $AP(1 - \cos(t))$. Of course, for this displacement of A, P can have swung from one side of AC to the other, giving a total of $2AP \sin(t)$. The enhancement achieved is then

$$\frac{2\sin(t)}{1 - \cos(t)} = \frac{2\sin(t)}{\text{versin}(t)},$$

using a slightly old-fashioned trigonometrical term 'versed sine', equal to one minus the cosine. Putting some numbers in shows that for $t = \pm19°$ one foot of straight motion of P can be had from only about one inch of guide, a ratio of 12:1.

Scott Russell's linkage occurs in all sorts of unexpected places once you start looking for it. Indeed, you may be carrying around an example in your car in the form of a car jack. In one popular design, the points C and P are connected by a long, threaded screw. The load of the car comes down on A. By rotating the screw, P is pulled closer to C, lifting the car. Here the mechanical advantage works in the opposite direction: a large horizontal movement of P causes a small vertical movement of A, making it easier to lift the car. A model of this linkage is shown in plate 7.

A slightly cautionary note here: suppose AP is a ladder resting against the side of a house, then the best way to stop it sliding down is to fix P. No matter how strong any rope securing B to C may be it is entirely useless.

Going back to the Scott Russell linkage we can see that the short guide GG could, in principle at least, be replaced by any of the linkages already described in this chapter. A Watt linkage for example gives a very close approximation to a straight line near to its centre of movement and would be quite suitable. Chebyshev experimented with arrangements, including those with his linkages. An engine using this arrangement was displayed at the Vienna Exhibition in 1873. Comments on this engine were reported in *Engineering* on 3 October (p. 284).

> The motion is of little or no practical use, for we can scarcely imagine circumstances under which it would be more advantageous to use such a complicated system of levers, with so many joints to be lubricated and so many pins to wear, than a solid guide of some kind; but at the same time the arrangement is very ingenious and in this respect reflects great credit on its designer.

So, by the time of Scott Russell, in practice any gains that could come from replacing the guide by another Scott Russell linkage or other arrangement were not really worthwhile. There is one exception, though, that found application in medium-sized steam engines. It is based on the fact that a short arc of a large-radius circle does itself not deviate greatly from a straight line (shown in figure 2.23). In the form used in engines, D is fixed and a new link AD added. Point P guides the piston rod as it is driven by a vertical cylinder. It is known as the *Grasshopper* link because of the way link AD appears to vibrate when the engine is running. The place for this linkage should really be section 2.1 as P cannot move in a true straight line, but it is more appropriate to include it here with Scott Russell. A model of this linkage is shown in plate 8. Notice the similarity with the right-hand side of figure 2.13.

2.5   Other Ways to Draw a Straight Line

There are a variety of other ways in which a truly straight line can be drawn which do not involve link works. Just two are given here, the first of which involves one circle rolling around inside

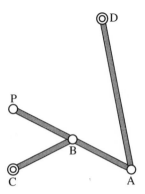

Figure 2.23. A Grasshopper linkage.

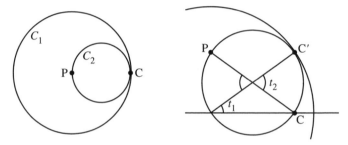

Figure 2.24. Rolling C_2 inside C_1.

another. The second relies on two identical parabolas rolling around each other.

Begin with two circles: C_1 of radius r_1 and C_2 which has half the radius, so $2r_2 = r_1$. We roll the smaller circle around *inside* the larger one. Points on the perimeter of the smaller circle will be found to move in perfect straight lines. We begin in a configuration in which both horizontal diameters coincide. On the small circle mark two points: C where C_1 and C_2 touch, and P the point on the perimeter of C_2 which is also the centre of C_1. Note that C and P lie on the horizontal diameter of C_2. This is shown in the left-hand diagram of figure 2.24. Next we assume that C_2 has rolled around the inside of C_1, and this is shown, enlarged, in the right-hand diagram, so that the line which joins the centres of C_1 and C_2 makes an angle of t_1 with the horizontal.

We now have a new contact point, C′, between C_1 and C_2, and we consider the angle t_2 between the two radii in C_2 connecting the horizontal to C'. As we have assumed that there is no slipping between the two circles C_1 and C_2 as they roll, the *arc length* from C to C' on C_1 must equal that from C to C' on C_2. Arc length is the product of radius and angle, $l = tr$ (measured in radians of course). Since $r_1 = 2r_2$ we have that

$$r_1 t_1 = l = r_2 t_2,$$

or, since $2r_2 = r_1$,

$$2r_2 t_1 = r_2 t_2,$$

so that $t_2 = 2t_1$.

From this it is clear in the diagram, and can easily be confirmed by elementary geometry, that the point C remains on the horizontal diameter of C_1. Since t_1 is arbitrary in this argument it follows that the path of C is a horizontal straight line. Indeed, an identical argument shows that the point P moves along a vertical straight line, and indeed each individual point on C_2 moves in a straight line. This observation has been used with a small cog wheel inside a larger one to generate straight-line motion. Such a contrivance is that patented by James White in 1801 (see figure 2.25). One problem with this mechanism is the great strain on the central bearing, and it was not particularly widely used.

The next method for drawing a straight line is, admittedly, somewhat contrived. Nevertheless it does illustrate one property of the parabola that we learned how to cut out in section 1.4. There we gave an algebraic equation of the parabola in the form $y = x^2/(4a)$. First we consider the parabola from a geometric, rather than an algebraic, point of view. Let us take a point P on the parabola, which of course has coordinates

$$P = \left(x, \frac{x^2}{4a}\right).$$

Again, we use the Pythagorean theorem to find the distance of this point from the coordinate $(0, a)$. This turns out to be

$$\frac{x^2}{4a} + a,$$

Figure 2.25. James White's parallel mechanism.

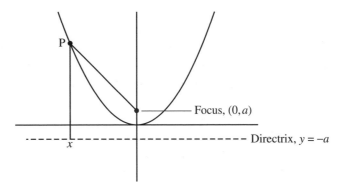

Figure 2.26. The focus–directrix approach to the parabola.

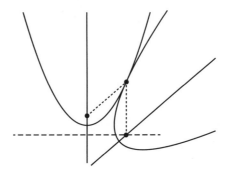

Figure 2.27. Two identical parabolas.

which is precisely the distance from P to the horizontal line $y =$ $-a$. Hence, we can show that the parabola is the locus of points that are the same distance from a given fixed point $(0, a)$ as they are, perpendicularly, from a line $y = -a$. The fixed point is called the *focus* of the parabola, and the line the *directrix* (see figure 2.26). Indeed, the parabola is more traditionally defined as *the locus of a point which moves so that its distance from a fixed point is equal to its perpendicular distance from a fixed straight line*. From this its equation is derived. This explains the slightly strange looking factor of $1/(4a)$ that we gave in the defining equation of the parabola.

What we now do is take two identical parabolas, one of which we assume is fixed. The second is made to roll around the first, with the condition that it does not slip. This configuration is shown in figure 2.27. There is a natural symmetry between the two parabolas, with the focus of one always lying on the directrix of the other. This means that if we place a pen on the focus of the moving parabola, it will draw a straight line as one parabola rolls around the other.

We have experience of trying to draw a line in this manner using the two parabolic plywood templates shown in plate 9. It is extremely difficult to do and is at least a three-hand job.

Chapter 3

FOUR-BAR VARIATIONS

It is quite conceivable that the whole universe may consti-
tute one great linkage, that is, a system of points bound to
maintain invariable distances, certain of them from certain
others, and that the law of gravitation and similar physical
rules for reading off natural phenomena may be the consequences of this condition of things. If the Cosmic linkage
is of the kind I have called complete, then determinism is
the law of Nature; but, if there be more than one degree of
liberty in the system, there will be room reserved for the
play of free-will.

Sylvester (1875b)

Linkages are such an important topic that we cannot restrict our
attention solely to drawing a straight line. The simplest linkage
mechanism contains only three obvious bars, and a fourth if you
count the fixed base. These found an application in generating
an approximate straight line. Of course, applications of four-
bar variations are not restricted to generation of approximate
straight lines. We shall examine these linkages in more detail and
consider the curves that four-bar linkages actually generate.

It is not possible to attempt to list all the other uses here,
so instead we shall take a small sample for the purposes of
illustration and suggest that you look around to see how often
they are used. Sometimes they are not obviously linkages, as
one or more of the links may form part of the whole structure,
but nevertheless they are there. Look at pushchairs, long-handle
loppers for pruning trees and much of the machinery used on
large building sites. Indeed the modern area of robotics relies
heavily on linkages and while we have concentrated here on
some historical uses, this is a field which is very much alive.

Figure 3.1. Four bars in the form of a parallelogram.

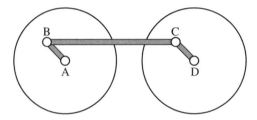

Figure 3.2. The coupling rod of a steam engine.

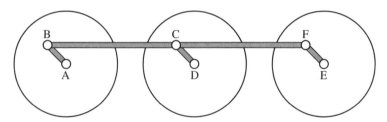

Figure 3.3. Coupling six wheels of a locomotive.

As a general comment on all linkages, and we can take Watt's as an example, it may not always be apparent that the fixed link AD is itself a link. These fixed pivots could be on a base board as in figure 2.6, or fixed to the engine house of a beam engine, or on the frame of the engine itself. Nevertheless, they act as links. We start with a parallelogram where AD = BC and AB = CD (see figure 3.1). The parallel rule used for navigation is a clear example because each member looks like a link. AD and BC are the rules and the shorter sides are plainly links. The word 'parallel' in this context signifies that the outer edges of the rules remain parallel however widely the rule is opened. This contrasts with the same word being used by early engineers to signify movement along straight parallel lines some prescribed distance apart.

A far less obvious linkage could be seen almost anywhere on the United Kingdom's steam railway system until the early

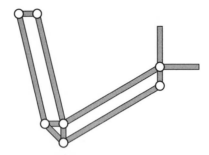

Figure 3.4. Parallelograms used on a drawing board.

1960s, and may still be seen on preserved locomotives. It is the coupling rod used to transmit power from the driven axle to other axles to improve the performance of the locomotive. The coupling rod is the one element of the parallelogram which looks like a link; the other longer link is the frame of the locomotive itself, and AB and CD are parts of the hubs of the driving wheels. Figure 3.2 shows a 'four coupled' locomotive, and, as an aside, the coupling rods on each side of a locomotive would normally be 90° out of phase to avoid problems when starting on dead centres, i.e. when A, B, C and D lie in a straight line. It was common to have six coupled wheels on locomotives and power was transmitted by two parallelograms joined end to end as shown in figure 3.3. So far it has been possible to designate one or two of the pivots as being fixed, and in this instance the fixed points are obviously the axles A, D and E. However, this is not actually practicable as all axles are sprung, giving each axle some degree of vertical movement to allow for imperfections in the track. This implies that BCF cannot be a single link with pivot holes at B, C and F, as it would then be subjected to bending stresses. Instead the rod is made of two sections, often sharing a pivot at C.

Two parallelograms can be joined in a slightly different way to form a linkage commonly used on drawing boards. Two rules were joined to make an L-shape, the purpose of the linkage being to allow this form of square to be moved over the board preserving its orientation, so the horizontal leg is always parallel to the base of the board with the other leg square to it (see figure 3.4).

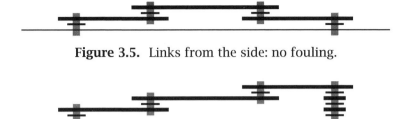

Figure 3.5. Links from the side: no fouling.

Figure 3.6. Links from the side: extra vertical space to avoid fouling.

3.1 Making Linkages

The main difficulty in making a model linkage is deciding how the separate links should be pivoted, and here we offer some suggestions. If you have a fully equipped workshop with a lathe and a vertical mill, you will not need any advice on how to use them to produce a properly engineered metal linkage—but do not dismiss the rest of this section just yet. However you make a linkage, of whatever material, it is always a great help if you start by making a mockup using strips of cardboard, plywood or Meccano. It does not matter how crude it is or how well it works as its purpose is to find how the links should be arranged in the plane so that they do not foul each other unnecessarily. As an example we consider Watt's linkage, and its more generalized form, where the distance between fixed pivots can be varied. The long arms of the Watt's linkage in figure 2.6 are designed to move only some 20° either way from their central position, and the links can therefore be positioned as shown in figure 3.5.

Note the provision of spacers/washers. This arrangement would not be suitable for the general linkage as the long arms would foul each other when the fixed pivots are arranged close together, so a better scheme is shown in figure 3.6.

In any linkage the crossing of individual links cannot be avoided. So in general, it is not possible to draw a complete curve: narrow links are a real advantage. A simple method we have used to make linkages is based on press studs and the wooden sticks or spatulas that are often provided with ice cream or in cafes and coffee bars in place of teaspoons. Press studs are remarkably good pivots as they turn freely, yet they have little play and

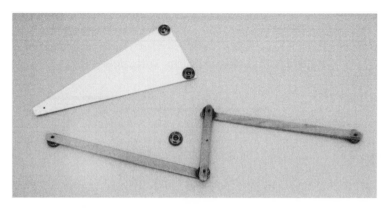

Figure 3.7. A simple model of linkages.

can easily be glued to the wood with an epoxy resin. They can
also be taken apart and the links reused. Haberdashery shops
and market stalls offer a wide range of sizes and designs, and
the best for our purposes are those with raised centres on the
base of the faces. The reason for this is that whenever a centre
is marked on a link it is difficult to glue the stud in position
accurately. It is better to define the centre with a small hole, say
1.5 mm in diameter, so as to provide a definite location for the
stud. It also helps if the hole goes through the wood to provide a
definite centre for another pivot or as an aid to further marking
out. Glue the studs separately before joining the links. For those
linkages where the tracing point is at a pivot point, Peaucellier's
linkage for example, a different technique is needed. The hole in
the two links should be just big enough to take the capillary tube
nib of a drawing pen, and the links should be held together with
a paper clip glued in position. It looks somewhat inelegant, but it
works. Photographic mount card is also a very useful material for
a base board, and also for triangular sections such as that which
occurs in Roberts's linkage. An example is shown in figure 3.7,
which allows Watt's, Chebyshev's and Roberts's linkages to be
demonstrated. This requires only two or three wooden sticks,
seven press studs and one sheet of card.

Metal linkages can be made without elaborate workshop facilities by using aluminium strips with pop rivets as the pivots
and washers to act as spacers. The idea is to start to tighten the

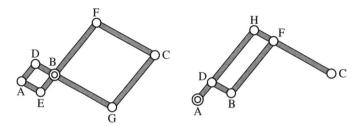

Figure 3.8. Forms of the pantograph.

rivets so that the links can turn freely and there is no risk of them separating. Then cut off the wire piece that would usually have come out if the joint had been properly tightened. Care is needed, however, to avoid sharp edges.

You may have other ideas for easy ways of making linkages— our only comment is that much more is gained by making a model linkage than static sketches.

3.2 The Pantograph

Another important simple linkage is a drawing instrument that is known as the pantograph and is used to enlarge or reduce drawings. Take a careful look at some of the photographs of linkages with engraved brass labels: the one in figure 2.14 for example. The engraving machine used stencils some eight times the size of the lettering and the reduction was achieved by means of a pantograph.

Its simplest form consists of the two rhombi shown on the left of figure 3.8. Imagine the point B fixed to the table. If the point C traces around a shape in the plane, then since ADBE has one-third of the dimensions of BFCG, the point A will trace around an identical shape, one-third the size and upside down. By extending the links AD and FC to intersect at H and then removing points E and G we obtain the usual form of the panto-graph shown on the right of figure 3.8. In this new configuration the points all move as before. It is more useful to obtain a copy of a figure which is the same way up as the original. This can be achieved with the linkage shown, where instead of fixing B to the

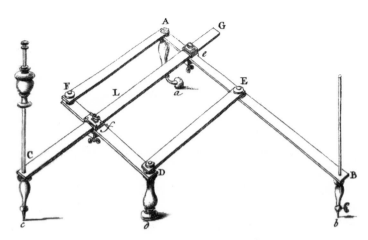

Figure 3.9. A pantograph.

table, A is fixed. Many other configurations are possible, such as that given in Stone (1753), which is reproduced in figure 3.9.

An important application of the pantograph occurs in what is known as an *engine indicator*. During the nineteenth century and for the first half of the last century reciprocating steam engines were the principal sources of power for ships, locomotives, works and factories of all kinds. It is essential to know how effectively and efficiently the high-pressure steam is being used to generate useful mechanical power, and this is what an indicator does. Its function is to record the pressure of steam in the cylinder at all stages of its stroke and on the return stroke too when the steam is being exhausted from the cylinder. An experienced engineer can see from this when the valve events are taking place—that is to say, when the high-pressure steam valve opens and closes and when the exhaust valve opens and closes, all in relation to the position of the piston in the cylinder.

The first indicator was also an invention of James Watt, and an early schematic is shown in figure 3.10(a). The heart of any indicator is a small cylinder and piston at P, connected to the end of the engine cylinder, closed by a spring-loaded piston. The strength of the spring is chosen to meet the supply pressure of the steam and its piston moves up and down in response to the pressure in the engine. This causes the point R, at which is a pencil, to move up and down. In addition, a string moves

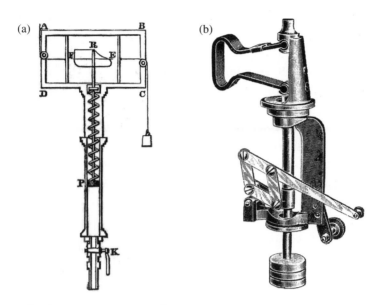

Figure 3.10. Steam engine indicators: (a) James Watt's; (b) Elliott Bros.

the frame horizontally in time with the stroke of the engine. The result is a closed curve that gives the pressure inside the cylinder at any given point of the engine's cycle. If correctly calibrated, the area enclosed by this curve is directly related to the work done by the engine in a single stroke, and from the stroke rate the *indicated power output* can readily be found.

In Watt's indicator above, the pencil is directly connected to the small cylinder. The design of the spring is such that the displacement of the piston is strictly proportional to pressure, and it is this that is recorded on the diagram. The actual movement is often too small to allow a legible diagram to be drawn so it must be magnified, hence the need for the guide linkage. The criteria for this linkage are more exacting than any we have mentioned before. First, the movement of the pencil must always be proportional to the movement of the piston: clearly a criterion not met by a Scott Russell linkage. Second, it must be light, with little friction, because the whole curve is drawn in one stroke of the piston. Many designs were on the market from different makers of indicators and in some, such as that of the

Elliott Brothers (figure 3.10(b)), a pantograph was used. Notice the external spring in this design, which is not affected by the high-temperature steam.

So far we have established a linear scale for the pressure axis— the corresponding problem for the stroke axis consists of reducing the displacement over a stroke to a dimension suitable for indicator cards, say around 100 mm. Again, many options were available, the pantograph amongst them.

In addition to its diagnostic uses, such as spotting leaking valves and leakage around the piston, the area of the diagram shows the power developed in the cylinder. From this the mean pressure in the cylinder can be calculated. This leads to the easily remembered formula

$$\text{power} = PLAN,$$

where

P = mean pressure,
L = length of stroke,
A = area of piston,
N = rotational speed for a single-acting engine
        (or $2N$ for a double-acting engine),

all taken in appropriate units.

Indicator diagrams were drawn by their thousands each year when steam engines were the prime movers and this goes to explain why planimeters, the topic of chapter 8, were, and still are, so important.

We would also like to mention here, not so much for its usefulness but for its delightful name, the *skew pantigraph* (Sylvester 1875b). This can be used for copying drawings with the copy turned through some angle from the original and also rescaled.

3.3   The Crossed Parallelogram

If the parallelogram is crossed, a new range of applications is possible. The Hart inversor already mentioned can form the

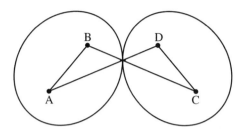

Figure 3.11. A crossed parallelogram.

basis of an exact straight-line linkage. Another application of a crossed parallelogram is in the steering mechanism of a cart. If the front and back wheels are both steerable, and connected by a crossed parallelogram, then the turning circle of the cart is vastly reduced. In fact, the crossed parallelogram has some remarkable properties. The first remarkable property is that in all positions, the angles at A and C (see figure 3.11) are equal, as are the angles at B and D. We shall assume that A and B are fixed and that C and D move. The question to resolve is the path of the hypothetical point P, not marked on the diagram, where the lines AD and BC cross. We note by symmetry that the distances APB and CPD are equal in any position of the configuration and are hence constant. Imagine a piece of string of fixed length from A to P and back to B. The string constrains the position of P—you may be familiar with the string-and-pin method for drawing an ellipse. Indeed, if we view A and B as the foci, then the locus of P will be an ellipse. We shall leave the detailed examination of this topic until section 13.3, where it is compared with a variety of other methods. Notice, however, that it is clear how to find the tangent to the ellipse from this linkage. Although it would be hard to construct a device for drawing an ellipse from this idea, Artobolevsky (1976, volume 2, p. 93) does indeed propose such a linkage. Since C and D can be thought of as being the foci of another identical ellipse, this linkage configuration can be used as the basis for elliptical gears.

Elliptical gears can be used to produce variation of speed between given limits. This has been used as a quick return motion, for example, and also in pumps, printing presses and other applications in which most of the work is done in one direction and it is important to return to the original position as quickly as possible. However, such mechanisms are not common. Creating circular gears is a difficult and specialist topic in itself, and the difficulty in measuring the perimeter of an ellipse adds to the complexities.

> The practical uses of the elliptical gear are endless, and it would be in greater use and favor, if it were not for the fact that its production, by the means ordinarily in use for that purpose, is as difficult and costly as the resulting gear is unsatisfactory.
>
> Grant (1907)

3.4   Four-Bar Linkages

In chapter 2 we looked in detail at four-bar linkages from the point of view of the generation of approximate straight lines. In fact, such linkages produce a variety of interesting curves. One way to see these is to use a model in which the distance between the fixed axes is adjustable. For example, in the models in plate 10 the longer arms are 5 units long and the short middle arm is 2 units long. In the positions shown the curve is close to Watt's linkage and exactly equal to Chebyshev's. All the intermediate positions are possible by moving the arms.

Taking an approach using only elementary trigonometry and algebra does not yield a simple description of the path of the point on the linkage—in this section we shall provide a formula. In particular we shall take the four-bar linkage ABCD, where $A = (x_1, y_1)$, $B = (x_2, y_2)$, $C = (x_3, y_3)$ and $D = (x_4, y_4)$. Simple applications of the Pythagorean theorem arrive at a system of equations such as

$$(x_1 - x_2)^2 + (y_1 - y_2)^2 = (r_1)^2,$$

where r_1 is the length of AB.

If we take $P = (x, y)$, and assume that P lies on the line BC, or on an extension of this line, then we arrive at the following weighted average for the position of P on the line BC, for some p_1 and p_2:

$$x = p_1 x_2 + p_2 x_3 \quad \text{and} \quad y = p_1 y_2 + p_2 y_3.$$

We then have a system of four quadratic equations and two linear equations. The prospect of solving such a system to obtain the position of P relative to the two fixed points A and D appears to be hopeless. However, with the aid of a modern technique, known as Gröbner bases, it is possible to solve systems such as this exactly. For the purposes of comparison with chapter 2 we shall assume that AB = CD = 5 and that CB = 2. If we assume that P is at the centre of CB, then $p_1 = p_2 = \frac{1}{2}$. However, we shall position A = $(-r, 0)$ and D = $(r, 0)$, so that for $r = 2$ we have Chebyshev's crossed linkage, and for $r = \sqrt{24} \approx 4.898$ we have Watt's linkage. Solving the resulting systems of equations gives the path of P as

$$r^4(y^2 + x^2) + 2r^2(y^4 - 26y^2 - x^4 + 24x^2)$$
$$+ (y^2 + x^2 - 24)^2(y^2 + x^2) = 0.$$

Notice that for each particular r this gives not an explicit equation for y in terms of x, but rather an implicit relation between x and y.

   If we solve this equation from a purely algebraic point of view, using numerical techniques if necessary, then we see that the curve for Chebyshev's crossed linkage has two disconnected curves, not just the single portion shown in plate 2. Of course, this is because continuous movement of a mechanism might not allow all parts of the feasible curve to be reached. Usually, whenever we have the intersection of two circles, or a circle and a line, there are two equally valid solutions. Hence, from a mechanical point of view this results in ambiguity as to which solution should be chosen. In a physical mechanism such ambiguity is usually, but perhaps subconsciously, designed out. When it exists it can give rise to very interesting dynamics, an example of which we will examine in section 3.7.

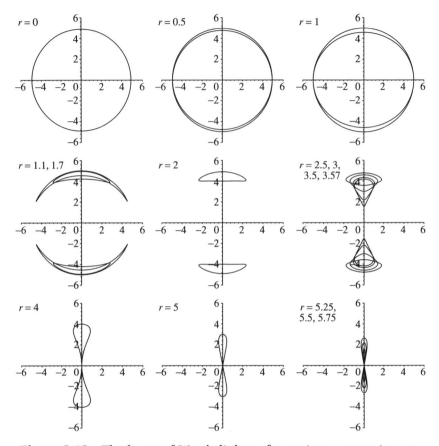

Figure 3.12. The locus of Watt's linkage for various separations r.

Figure 3.12 shows the variety of curves that can be generated by changing the distances between the fixed pivots shown in plate 10. They range from a point where all the links are in line with AD = 12, via Watt and Chebyshev linkages, to a circle where A and D coincide.

One particularly famous curve is known as the lemniscate of Bernoulli, shown in figure 3.13, which can also be drawn using a crossed parallelogram. If we take

$$AB = CD = 1,$$
$$BC = AD = \sqrt{2},$$

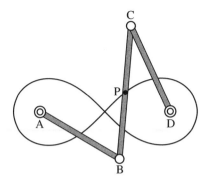

Figure 3.13. The lemniscate of Bernoulli.

with A and B as the fixed points

$$A = \left(-\tfrac{1}{\sqrt{2}}, 0 \right), \qquad B = \left(\tfrac{1}{\sqrt{2}}, 0 \right),$$

then the path described by P is

$$(x^2 + y^2)^2 = a^2(x^2 - y^2),$$

where a is a scale factor. The curve gets its name from the Greek *lemniskos*, a bandage or ribbon, via Latin.

3.5 The Triple Generation Theorem

We conclude our work on four-bar linkages with a most remarkable result known as the triple generation theorem. You may recall that in section 2.1 we found two arrangements of linkages that resulted in one path for the point P. In this section we find *three* linkages that all result in generating the same path. It is deceptively simple to do this, and to start we draw an arbitrary triangle ABC and select any internal point P. If we draw lines through P parallel to the sides of ABC then we can create the three general four-bar mechanisms shown in figure 3.14. Here, just as with Roberts's approximate-straight-line mechanism, the point P does not lie on the middle linkage.

Now, if we fix pivots A and B closer together to allow the point P to move, then C remains in one place as P moves. Alternatively, the three general four-bar mechanisms each constrain P to move along the same locus.

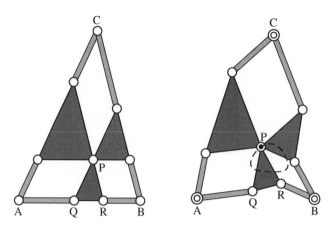

Figure 3.14. The triple generation theorem.

The mathematician will probably be wondering if an arrangement of linkages can be found which will generate *any* given curve. A paper by Kempe (1876) showed that for *any algebraic curve* there exists a linkwork which will generate it. The original paper is somewhat condensed and Yates (1949), who works up to this result through some 200 pages of carefully structured exercises on the subject, is probably more enjoyable to read.

3.6   How to Draw a Big Circle

It might seem strange to consider here how to draw a circle, since a rudimentary pair of compasses can be fashioned from almost any rigid object. However, drawing the arc of a circle with a very *large* radius is quite problematic in practice. Not only will the centre be a long way away, but constructing and moving a large rigid object is never easy. Draftsmen used a collection of template arcs of prescribed radii, known as *railway curves*. Although this is a practical solution when the radii are known, it is unsatisfactorily restrictive. While in chapter 1 we suggested constraining a wedge between two pins, this is hardly going to provide a technique sufficiently accurate for engineering purposes. Furthermore, in keeping with the rest of this chapter we would prefer to use rigid linkwork to constrain the motion.

Perhaps it is not surprising that Watt's linkage can generate an approximate circular arc of large radius, and indeed this arrangement has been applied to cutting and polishing large stone blocks by Keady, Scales and Fitz-Gerald (2000). However, Peaucellier's cell, which gave an exact straight-line motion, can easily be adapted to generate an exact circular arc of very large radius. The resulting device is compact and it is not necessary to be able to physically get to the centre of the circle.

To see why this is the case, we recall the fundamental relationship between the points O, C and P in figure 2.16. Peaucellier's cell is an inversor, which is to say that $OC \cdot OP = k^2$. We shall forget temporarily about linkages and just concentrate on this mathematical relationship between points in the plane. We shall assume that a point C is a distance r from the origin and that the line OC makes an angle t with the horizontal. Then the point P is a distance k^2/r from the origin, and OP is also at an angle t with the horizontal. The relation $OP = k^2/OC$ gives rise to the description of P as the inversion of C in the circle of radius k.

In general, if point C moves along a line through the origin, then the angle remains constant. Hence P will also move along the same line. If C moves along a circle through the origin, as we have already seen, P moves along a straight line. Conversely, if C moves along a straight line, P follows a circular arc, which is part of a circle through the origin. This might allow us to draw arcs of large circles, but we have already seen that constraining a point to move in a straight line is problematic. The last case is when C moves on a circle not through the origin. Here P also moves along a circle. Proving these statements using elementary trigonometry is not a pleasant task—a much more aesthetically appealing approach requires the slightly more sophisticated machinery of complex analysis, which we choose not to examine in detail in this book. See instead the elegant book of Needham (1997).

For us, if we change the length of the distance OQ in figure 2.16, then the circle on which C is constrained to move will no longer pass through O. As a result P will now move along a circular arc, and it is the radius and centre of this we now find using elementary methods. Notice at the top of plate 4 that the additional holes allow exactly this situation to be illustrated. We shall not find the extent of the circular arc which a configuration of linkages can generate. Although this is possible, the resulting formulae are not particularly enlightening.

We appeal to symmetry to argue that the centre of the circle along which P moves must lie on the x-axis. Although the inverse of the centre of the circle does not correspond with the centre of the inverted circle, we can find the centre by taking the average of two points on a diameter. Adopting the same notation as in figure 2.16, we now assume that $OQ = l_3$ and $QC = l_4$, where now it is no longer the case that all these links are of equal length. Assuming that P moves on a circle with centre at centre $R = (l_5, 0)$ and radius l_6 we have the two potential situations. The first assumes that $l_3 > l_4$, shown at the top of figure 3.15, and the second assumes that $l_4 > l_3$, shown at the bottom of the figure.

For the case in which $l_3 > l_4$ we have that

$$l_5 + l_6 = \frac{k^2}{l_3 + l_4} \quad \text{and} \quad l_5 - l_6 = \frac{k^2}{l_3 - l_4}.$$

Solving these equations gives

$$l_5 = k^2 \frac{l_3}{l_3^2 - l_4^2} \quad \text{and} \quad l_6 = k^2 \frac{l_4}{l_3^2 - l_4^2}.$$

The case in which $l_3 < l_4$ is very similar.

Notice that if l_3 and l_4 are very similar then the radius, l_6, is very large indeed and the centre correspondingly far away. If $l_3 = l_4$ then we would need to divide by zero, giving an infinite radius, which is forbidden. In this case we already know that we have a straight line, although we shall again appeal to the fiction that a straight line is a circle of infinite radius in section 8.5.

3.7   Chebyshev's Paradoxical Mechanism

We end this chapter with a most intriguing and remarkable mechanism which Artobolevsky (1976, section 601) describes as Chebyshev's paradoxical mechanism. A model of this mechanism is shown in plate 11. It is described as 'paradoxical' since

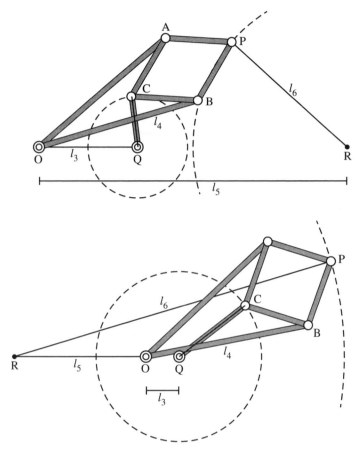

Figure 3.15. Using Peaucellier's linkage to draw a large circle.

one revolution of the crank handle corresponds to two revolutions of the flywheel in the same direction, or to four revolutions in the opposite direction. This mechanism really is worth constructing, if only to confound your friends and colleagues. A schematic of the mechanism is shown in figure 3.16.

The lengths of the linkages must satisfy

$$AB = CB = MB = 1,$$

$$FC = 1.387, \quad EA = 0.557, \quad DM = 0.584,$$

$$CE = 1.324, \quad FD = 0.123.$$

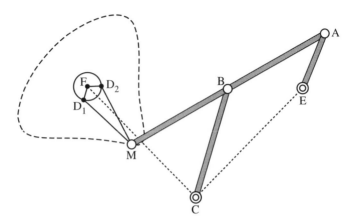

Figure 3.16. Schematic of Chebyshev's
six-bar paradoxical mechanism.

Furthermore, the angle FCE, shown as a dotted line in the diagram, must be a right angle. The crank handle is at A, and the flywheel is pivoted at F. In this diagram, the links BC, AE and AM form a four-bar linkage, which is very similar to the alternative form of Chebyshev's approximate-straight-line motion shown in figure 2.12. The locus of the point M as the point A is moved around a circle is shown as a dashed line in the diagram.

There are two other links to complete Chebyshev's paradoxical mechanism: these join F to a point D, and D to M. These are shown as lines, rather than links, in the diagram. Since the position of the point D is given as the intersection of two circles, one centred at F of radius FD = 0.123 and one centred at M of radius DM = 0.584, there are *two choices* for the position of the point D, shown as D_1 and D_2 on the diagram. It is precisely the ambiguity between these two choices that gives rise to the apparently paradoxical behaviour of the flywheel at F when A is rotated.

What is so unusual in this mechanism is the ability of the linkages to flip from one configuration to the other. In most linkage mechanisms such ambiguity is implicitly, or explicitly, designed out so that only one choice for the mathematical solution can give a physical configuration.

Chapter 4

BUILDING THE WORLD'S FIRST RULER

Three grains of barley, dry and round, make an inch.

Proclamation of King Edward II, 1324

In chapter 2 we considered how to draw a straight line. This turned out to be a little more difficult than one might assume. What we shall do in this chapter is turn this straight edge into a graduated ruler, so that we can actually begin to measure, and in order for this ruler to be useful we should ensure that the graduations are equally spaced. Just as we cannot use a straight edge to draw the first straight line, so we shall not be able to use an existing ruler to mark out the graduations on our first ruler. Instead, we shall use pure geometry to construct the positions of the markings. This will also prove to be tricky as we must follow Euclid's example and restrict ourselves to having available only the straight edge we have constructed and a pair of compasses. Indeed, as we shall discover there are some lengths that we simply cannot mark out exactly on our ruler in this way.

Before we can enjoy the purely geometrical recreation of dividing the unit length, however, we must decide what this unit should be, and since this is the first ever ruler we have complete freedom of choice. We are not concerned here with marking a straight edge to an accepted standard but rather with graduating an accurate scale to the greatest precision within our powers. As we show, practical difficulties have arisen because of the proliferation of a wide range of standards, thus sacrificing interchangeability of components from different manufacturers.

4.1 Standards of Length

Standards of length have a history that is almost as old as that of mankind. Many of the earliest standards refer to human body parts such as a 'hand' or a 'foot'. However convenient these were, with increasing trade it became apparent that these were arbitrary. Indeed, there was quite a wide range even in a unit such as the foot, between the Pythic foot at $9\frac{3}{4}$ (modern) inches and the foot in Geneva at some 19 in. We now have independent standards of length but it is still surprisingly useful to know the width of your thumb or the length of your span. To give you some idea of the kinds of accuracy necessary for engineering, the following quote of Bond (1887, p. 112) is instructive, since it shows that over 100 years ago engineers were taking very fine tolerances for granted.

> The arm of King Henry the First, or the barley corn, though possibly furnishing a standard good enough at that time, would hardly satisfy the requirements of our modern mechanics or tool makers, who work very often within the limit of a thousandth of an inch, and even *one-tenth* of this apparently minute quantity, with surprising unconcern and no less accuracy.

In manufacturing interchangeable screw threads, for example, it is vital to be able to implement standards accurately. Assume that we have independently manufactured nuts and bolts. In order for these to work together the profile of each thread must be identical. Not only in the shape of the profile, but also in the size. In particular, any differences in the *pitch* of the thread, the distance between corresponding peaks in the screw, will be cumulative so that unless made accurately the nut and bolt will lock and cease to function correctly.

Hence, standards of length have to be very accurate and yet practical and reproducible. More importantly, the development of a standard has to be independent of the thing to which it refers. Hence in 1671, Picard used time as the basis of his standard yard. He proposed to take a pendulum that has a period with a duration of two seconds. The length of such a pendulum would provide the basic standard length. However, the Earth is an oblate spheroid rather than a true sphere and a given pendulum will oscillate more rapidly at the poles than at the Equator. Conversely, a pendulum made to beat seconds at 4° 56′ N (i.e. near the Equator) will differ by over one-tenth of an inch in length from that at 48° 50′ N (i.e. near Paris). Hence, if a pendulum is to be used it will have to be standardized in one place, or be subject to a correction factor.

In 1718, Jacques Cassini (1677–1756) proposed that a standard length should be $\frac{1}{6000}$ part of a minute of a degree of a great circle of the Earth. This corresponds approximately to a foot, although which great circle is taken does matter. As a result, in 1791 the metre length was standardized at 10^{-7} of the meridian through Paris by the French Academy of Science. Once this length had been established, a standard reference metre, called the *Mètre des Archives*, was made in a platinum–iridium alloy.

Pendulums, or fractions of an imaginary circle on the surface of the Earth, are certainly independent but they are not immediately practical. Indeed, before the twentieth century all such appeals to independent physical measures were not sufficiently reliable, and as a result metal bars became important reference objects. Such a bar was enacted to be a standard through law, rather than because of any intrinsic physical relationship. These are practical when they are available for reference, but they do have a number of drawbacks. One is their vulnerability to damage. Indeed, the British standard yard of George IV, which was enacted to become the standard on 1 May 1825, lasted only nine years before being damaged by the fires which destroyed the old Houses of Parliament on 16 October 1834. Metal is also subject to thermal expansion, so a bar only represents a standard length at a particular temperature. Picking a bar up in your hands for a few seconds imparts sufficient heat to cause havoc! These metal bars also flex, depending on how they are supported, and this can have a measurable effect on their length. Indeed, Bond (1887, p. 7) reports that the deviation in the length of one standard yard bar when supports are placed at the extreme ends was over one-thousandth of an inch: a length significant in many engineering applications. In the mid nineteenth century, it was also not known whether the alloys used changed length naturally over time, particularly in view of corrosion and other chemical reactions.

Even if we have a physical standard yard, there is the issue of what to do with it. Up until 1798 measures of length were transferred from the standard to a copy using beam compasses. Not only is the compass subject to flexure and thermal expansion but, worse still, the actual physical contact of compass points with the standard bar is bound to damage it sooner or later. Edward Troughton (1753–1835) was the first person to use optical instruments for this purpose instead. These optical techniques were developed throughout the nineteenth century, so that sophisticated machines were in use by the end of the century to work from a standard bar, subdivide the unit and transfer these measurements to gauges for use in a workshop. For example, the Rogers–Bond Universal Comparator shown in figure 4.1 is described by Bond (1887, p. 50) as follows.

> This apparatus is used, firstly to compare line-measures of length with attested copies of the standard bars of England and the United States; second, to sub-divide these line-measures into their aliquot parts, and to investigate and determine the errors, if any, of these subdivisions; and thirdly, to reduce these line-measures to end-measures for practical use in the shops.

Optics is currently used not only to measure and compare lengths, but also to define the standard. While the nineteenth-century scientists had little success in using an independent physical quantity, the metre, which is the base unit of length in the International System of Units, is now defined to be the distance travelled by light in a vacuum during one 299 792 458th of a second.

While all this seems to appeal only to a particular kind of pedant, it is actually something without which modern life could not begin to function. The ability, for example, to interchange different bolts with a given nut is something we take for granted. This applies to all other manufactured goods: the components

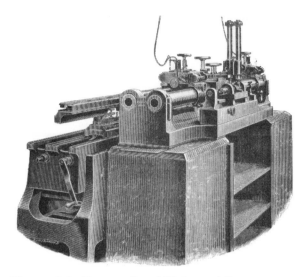

Figure 4.1. Rogers–Bond Universal Comparator.

in the hard drive of your computer, or the shape of the groove in the ring pull of an average drink can, for example. Both are made within tolerances that are hard to imagine. Indeed, one of the wonders of the modern world is the ability to mass produce *precision* items at minuscule cost.

4.2   Dividing the Unit by Geometry

Since we are trying to make the 'first' ruler, we shall abandon modern optical techniques and return to the kinds of pure geometrical constructions used in the eighteenth century and before. We shall further assume that we have agreed a unit, and marked it on what will become our ruler. What we must do now is consider how we might make a marked scale such as a ruler from scratch, using only the simplest drawing instruments.

Restricting oneself to using a straight edge and a pair of compasses is a classical topic of geometry. Euclid took this approach, not because they were the only instruments of his day, but rather because he wanted to construct his geometrical theory using a minimum of assumptions. That is to say, he wanted to assume only a small collection of self-evidently true *axioms* and derive, in a logically sound manner, the consequences of these. His assumptions were that it is possible to

1. draw a straight line from any point to any point,
2. extend a finite straight line indefinitely in a straight line,
3. draw a circle with any centre and any diameter.

Furthermore, he assumed that

4. all right angles are equal to one another, and
5. that if a straight line falling on two straight lines make the interior angles on the same side less than two right angles, the two straight lines, if extended indefinitely, meet on that side on which the angles are less than the two right angles.

This last assumption is sometimes known as the *parallel postulate* and has been a source of much discussion and confusion over the centuries. Notice, however, that the first three axioms correspond exactly with the operations that our ruler and compass will help us perform. These operations are one way of remembering Euclid's axioms. They can also be undertaken reasonably accurately by skilled hands.

To begin, we draw a straight line and arbitrarily mark two points on this line. We decide the distance between these will be our unit length: that is, a length we shall identify with the number 1. So, we now have our ruler with two points marked, nominally at 0 and 1. What else can we mark?

Certainly we can open our compasses to this distance and draw a circle centred at 1. This will intersect the line again at 0 and on the other side at another point, which of course we label 2. Repeating this we can easily mark off every whole number, both positive and negative, on our new ruler.

We can of course combine these basic operations to form a sequence. For example, imagine that we are given a line and any point (either on the line or not). We want to draw another line perpendicular to the first through the point. It is not difficult to show that this is possible using only a straight edge and compass. Specifically, take the line l and point P, as shown in figure 4.2. The task is to draw a line through P perpendicular to l. First draw a circle with centre P with a radius sufficiently

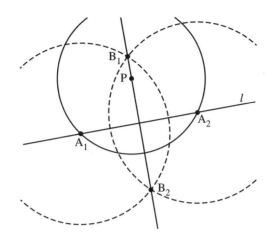

Figure 4.2. Constructing a perpendicular to
a line through a specified point.

large so that it intersects l twice. These intersection points are
A_1 and A_2. Open the compasses to a radius larger than half the
distance between A_1 and A_2. Next draw two circles, one centred
at A_1, the other at A_2. The points where these circles (the dashed
lines in figure 4.2) intersect are labelled B_1 and B_2. Connect B_1
and B_2 with a straight line. This will be perpendicular to l and
pass through P.

If two points are given, A_1 and A_2 say, the length corresponding to half the distance between these may be constructed in
exactly this way. Similarly we can divide by four, eight and so
on. This enables us to add $\frac{1}{2}$, $\frac{3}{2}$, $\frac{1}{4}$, etc., to our ruler.

The above examples are purely geometrical. We can also use
these tools to perform arithmetic by identifying the length of a
line segment with a number. In the next section we expand on
this in a more systematic way and ask which numbers are *constructible* by using repeated applications of the two basic tools.
That is to say we shall build the world's first marked ruler. In
the process it turns out that it is not possible to construct some
numbers. This leads on to three famous impossibility proofs. In
particular, using only a straight edge and compass, the following
turn out to be *impossible*.

1. Given a cube, construct another cube of double the volume.
   Known as 'duplicating the cube'.

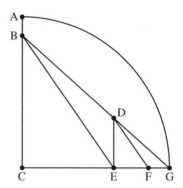

Figure 4.3. A geometric construction of $\frac{16}{113}$.

2. Given a circle, construct a square with the same area. Known as 'squaring the circle'.

3. Given an arbitrary angle, construct an angle one-third as large. Known as 'trisecting the angle'.

Before we do this, we digress and describe here an old Chinese method for constructing a line $\frac{355}{113}$ units long (a good approximation to π), as given by Gardner (1966). It is worth mentioning as an example of a Euclidean construction, and also to show how a fraction with the awkward looking denominator 113 can be drawn using only Euclidean means. Start with a quadrant ACG of unit radius, as shown in figure 4.3, with AB $= \frac{1}{8}$AC and join B and G. With DG measuring half a unit in length, drop a perpendicular to meet CG at E and now draw DF parallel to BE. FG is then $\frac{16}{113}$ units long.

The proof that FG $= \frac{16}{113}$ this assertion relies on the Pythagorean theorem as well as the properties of similar triangles, so we begin by noticing that

$$BG^2 = CG^2 + BC^2 = 1 + \left(\tfrac{7}{8}\right)^2.$$

Hence BG $= \frac{\sqrt{113}}{8}$. By the properties of similar triangles,

$$DE = \frac{BC \cdot DG}{BG} = \frac{7}{2\sqrt{113}}$$

and

$$EG = \tfrac{8}{7}DE = \frac{4}{\sqrt{113}}.$$

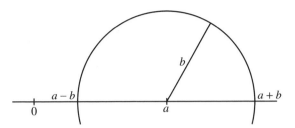

Figure 4.4. Constructing $a \pm b$.

CE $= 1 -$ EG and EF/DE $=$ CE/BC, from this we have

$$\text{EF} = \frac{7}{2\sqrt{113}}\left(1 - \frac{4}{\sqrt{113}}\right) \cdot \frac{8}{7} = \frac{4}{\sqrt{113}} - \frac{16}{113},$$

and hence FG $= \frac{16}{113}$ as claimed. We actually wanted to construct not $\frac{16}{113}$ but $\frac{355}{113}$ originally. To complete the construction, the unit radius is stepped off three times with the compasses to bring the length to $\frac{355}{113}$.

4.3 Building the World's First Ruler

We shall be a little more systematic in this section and we shall consider exactly which points we can mark on our ruler, and decide how to do this. We say that we can *construct the number* if we can construct the point on the ruler which is that length from the origin. What we shall show is that given any two previously constructed lengths a and b we may construct $a \pm b$, $a \times b$ and, provided that $b \neq 0$, $a \div b$.

From the previously constructed integers this allows us to construct all the rational numbers. But it also gives us a lot more, since every time we construct a new number, we can in a sense complete the number system by incorporating this with the existing collection of numbers using the same operations.

So to work. Constructing $a \pm b$ is very straightforward using the construction shown in figure 4.4. Simply open the compasses to a radius of b and draw a circle with centre a. The circle will intersect the line at the points $a + b$ and $a - b$. It really is as simple as that.

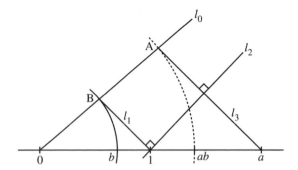

Figure 4.5. Constructing $a \times b$.

To construct $a \times b$, or more simply ab, from the two lengths a and b we draw a line and mark the positions of 0, 1, a and b as shown in figure 4.5. Note that the diagram illustrates the case $0 < b < 1 < a$, although this is unimportant. Mark any other line through 0, called l_0, and draw a circular arc with centre 0 and radius b. Where this arc intersects l_0, mark on the point B. Connect B and 1 with a straight line l_1. Mark the line, l_2 say, perpendicular to this through 1. Mark another line, this time l_3, through a and perpendicular to l_2. Where this intersects l_0 we mark the point A. If we choose we can mark the length 0A on the original line with a circular arc. In any case, the length 0A is indeed the required length $a \times b$. This may be seen since the two triangles aA0 and 1B0 are similar, so that

$$\frac{0A}{a} = \frac{b}{1}.$$

That is to say, $0A = ab$.

To construct $a \div b$ we refer to figure 4.6 and draw a straight line upon which we mark 0, 1, a and b. Again, we draw an arbitrary line l_0 through 0. Next, draw the circular arc through 0 of radius 1 and label the intersection of this arc with l_0 as B. Connect B to the point b. This line is l_1, and from this drop the perpendicular through a. Draw a further perpendicular to this new line through a and where this crosses l_0 we mark a point A. By drawing a circular arc back onto the baseline we have constructed the length $a \div b$. To see this we notice that the two triangles $0a$A and $0b$B

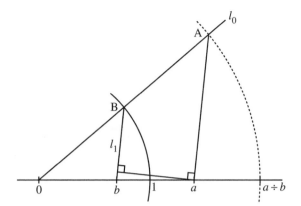

Figure 4.6. Constructing $a \div b$.

are similar. Therefore,

$$\frac{0A}{a} = \frac{1}{b} \quad \text{or} \quad \frac{0A}{1} = \frac{a}{b},$$

and therefore $0A = a \div b$.

Armed with the four arithmetic operations (addition, subtraction, multiplication and division) together with the two numbers 0 and 1 we may construct any *rational number* m/n, where m is an integer and n is any non-zero integer. Thus all rational lengths are possible. If you are not familiar with rational numbers, look ahead to chapter 9.

4.4 Ruler Markings

In section 9.1 we show that there are other lengths that are not rational numbers. An example is $\sqrt{2}$, although this length is easy to construct using a ruler and compass since we just take the diagonal of a unit square.

Hence, we may construct considerably more than just rational numbers m/n. In particular, given any positive length a we may construct $\sqrt{a}$ as follows. With reference to figure 4.7 construct the lengths 0, a, $a + 1$ and $\frac{1}{2}(a + 1)$ using the above constructions. Now draw the circle centred at $\frac{1}{2}(a + 1)$ of radius $\frac{1}{2}(a + 1)$. Draw a line through a perpendicular to the baseline and mark where this line intersects the circle, by A say. We are interested

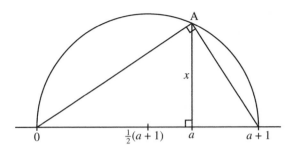

Figure 4.7. Constructing $\sqrt{a}$.

in the distance from A to a, which we call x. By theorem 1.1, the triangle through 0, A and $a + 1$ has a right angle at A. Therefore the two smaller triangles in figure 4.7 are similar. Using similarity we have

$$\frac{x}{1} = \frac{a}{x},$$

so that $x^2 = a$, from which we have $x = \sqrt{a}$.

So we can now add a mark for $\sqrt{a}$ on our ruler if there is already a mark for a. What else is possible? Can we mark $\sqrt[3]{a}$? What about π? In fact, can we mark *any* point we can think of, and if not can we decide exactly which numbers are constructible?

To answer this last question we now think of coordinates of points in the plane, rather than restricting our attention to our straight-line ruler. Given a point P_i, we think of this as having coordinates (x_i, y_i). So, for example, if the first two points we construct mark out our unit length on a horizontal line we think of these as having coordinates $(0,0)$ and $(1,0)$. Our straight edge and compass constructions allow us to generate points only from the intersections of

1. two lines,
2. a line and a circle,
3. two circles.

What we consider next is how to describe these possibilities using algebra.

First, assume we have two points P_1 and P_2 with coordinates (x_1, y_1) and (x_2, y_2). Take another arbitrary point on the line

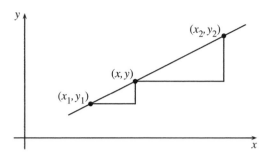

Figure 4.8. Points on a straight line.

connecting these, and call this (x, y) (see figure 4.8). By similar triangles,

$$\frac{x_1 - x}{x - x_2} = \frac{y - y_1}{y_2 - y}.$$

Rearranging this gives

$$x\underbrace{(y_1 - y_2)}_{:=a} + y\underbrace{(x_2 - x_1)}_{:=b} = \underbrace{x_2 y_1 - x_1 y_2}_{:=p}.$$

Since x_1, x_2, y_1 and y_2 are constant we can express this as

$$ax + by = p,$$

which is one form of the equation of a straight line.

Let us take four points $P_1, \ldots, P_4$ with coordinates (x_i, y_i). The intersection of the lines through P_1, P_2 and P_3, P_4 will be given by any solutions (x, y) to the simultaneous equations

$$x(y_1 - y_2) + y(x_2 - x_1) = x_2 y_1 - x_1 y_2,$$
$$x(y_3 - y_4) + y(x_4 - x_3) = x_4 y_3 - x_3 y_4.$$

Now, a solution may not exist (and indeed *does* not if the lines are parallel), but if a solution does exist finding it requires only the use of the mathematical operations of addition, subtraction, multiplication and division.

Next take a circle of radius r centred at P_3 (with coordinates (x_3, y_3)). The Pythagorean theorem tells us that points on the circle with coordinates (x, y) must satisfy the relationship

$$(x - x_3)^2 + (y - y_3)^2 = r^2.$$

When this circle and a line through P_1, P_2 meet, we have simultaneous equations:

$$x(y_1 - y_2) + y(x_2 - x_1) = x_2 y_1 - x_1 y_2,$$
$$(x - x_3)^2 + (y - y_3)^2 = r^2.$$

Again, there is no reason to suppose that the line and the circle do intersect. If they do, the coordinates x and y are related using only the operations of addition, subtraction, multiplication and division along with the extraction of square roots.

The case of the intersection of two circles is similar. The only mathematical operations needed to generate the coordinates are addition, subtraction, multiplication, division and the extraction of square roots. What we have done is turn a geometrical construction of lines and points into an algebraic discussion of what operations one is allowed to use to construct our numbers. The above discussion points the way to the following theorem, which characterizes precisely which numbers are constructible.

Theorem 4.1. *A number is constructible if and only if it may be obtained from the integers by repeated use of addition, subtraction, multiplication, division and the extraction of square roots.*

Proof. If a may be obtained from the integers by repeated use of addition, subtraction, multiplication, division and the extraction of square roots, then the above constructions show how we may construct the number with a straight edge and compass.

Conversely, let us assume that a number a is constructible. This means that there is a finite sequence of simple ruler and compass constructions from $(0,0)$ and $(1,0)$ that generate points P_1 and P_2, the distance between which is the length a. Each point generated in the construction occurs only at the intersection of two lines: a line and circle or two circles. From the discussion immediately preceding the theorem, we can see that the coordinates of such points can be obtained from previous coordinates using only addition, subtraction, multiplication, division and the extraction of square roots. Since we have a finite sequence of such points starting from $(0,0)$ and $(1,0)$, the coordinates of P_1, P_2 are obtained from the integers by repeated use of these operations. Since

$$a = \sqrt{(x_1 - x_2)^2 + (y_1 - y_2)^2}$$

it follows that a can be obtained using only addition, subtraction, multiplication, division and the extraction of square roots. □

Although the above is only a sketch of the proof of the converse, a more rigorous discussion can be found in, for example, Stewart (1989, theorem 5.2). What theorem 4.1 tells us is that many lengths are *impossible* to construct using only a straight edge and compass. However, given a number a it may not be easy to tell whether it can be expressed using the above operations. In fact, whether a is constructible depends on whether it occurs as the root of certain polynomials with rational coefficients. For example, the square root of a occurs as the root of the equation $x^2 - a = 0$. If a number a satisfies one such polynomial then it will satisfy many. What is important is the smallest degree of such a polynomial. It turns out that a is constructible if and only if the degree of the smallest polynomial is a power of 2, i.e. 2^n.

The cube root of 2, that is $\sqrt[3]{2}$, is not constructible since it satisfies the equation $x^3 - 2 = 0$ but no polynomial equation with rational coefficients of degree 1 or 2. This allows us to solve the following classical problem. Given a cube, with unit sides, is it possible to construct a cube with twice the volume? In order to do this we need to construct the length $\sqrt[3]{2}$. The fact that $\sqrt[3]{2}$ is not constructible shows that it is impossible to 'duplicate the cube'.

Numbers which do not occur as the root of any rational polynomial equation are known as *transcendental*. To conclude that π cannot be constructed using straight edge and compass we need to know that π cannot be constructed using the additional operation of taking square roots. In fact, the number π does not satisfy *any* rational polynomial and so cannot be constructible. The proof that π is transcendental is too long to repeat here, but can be found in, for example, Stewart (1989, chapter 6). Another example of a transcendental number is the base of the natural logarithms, e, which we shall encounter in chapter 8. Hence it is also impossible to construct this number.

The impossibility of constructing π solves the second classical problem, that of 'squaring the circle'. Given a circle of radius r, can we construct a square with the same area? Since the area of the circle is πr^2, we need to construct the length $\sqrt{\pi} r$. This is possible if and only if we can construct π. Since we cannot do this we may not square the circle.

These well-known impossibility proofs are encountered as part of most university undergraduate mathematics degrees. However, there are still those who seek in vain a geometrical construction that will square the circle, duplicate the cube or trisect the angle. Such people sometimes have a good grasp of geometry, and certainly great tenacity. Unfortunately, these attempts are bound to be futile, and the resulting lengthy calculations must inevitably contain at least one error. Professional mathematicians, including the authors, still occasionally receive unsolicited proofs of the impossible. The most recent bundle of papers one of us received arrived by airmail and included a 'proof', using exactly the ruler and compass constructions above, that

$$\pi = \frac{14 - \sqrt{2}}{4}.$$

Anyone with a pocket calculator can immediately see that

$$\pi \approx 3.141\,592\,6\ldots \neq 3.146\,446\,6\ldots,$$

which should be enough to suggest strongly that something is amiss. So what should the professional do? Clearly there is not sufficient time to wade through pages of nonsense, carefully correcting the inevitable mistakes. Since we know these constructions have been proved to be *impossible*, we know without reading the argument that it must be flawed. One option is to ignore such mail, but this may also be a serious error since G. H. Hardy discovered one of the greatest mathematicians, S. Ramanujan, on the strength of an unsolicited bundle of papers and helped obtain for him a fellowship at Cambridge. This is a famous story, told by Hardy in his *A Mathematician's Apology* (Hardy 1967). Furthermore, to ignore such correspondence confirms any suspicion the authors may have that the professional mathematical establishment is in conspiracy against them. Often this strengthens the resolve of the author who eventually publishes privately, as a newspaper advertisement or even, if the calculations are sufficiently extensive (unfortunately, to them at least, a synonym for 'important'), as a book. When such work is combined with a religious conviction, the result can be far from nourishing to the soul, as anyone who has tried to read such works as MacHuisdean (1937) can attest.

4.5   Reading Scales Accurately

While this chapter is mostly concerned with the problem of marking out scales accurately, an equally important problem is that of actually reading the scales themselves. Of course, a magnifying glass or microscope is invaluable here, and this is exactly what was used for precision work with standard lengths. However, there are also techniques which allow an extra decimal place of accuracy to be read directly with the naked eye. For example, while it is possible to read a normal rule to the nearest millimetre with the naked eye, there is no way to read a scale to one-tenth of a millimetre directly. Furthermore, marking a scale with ten marks in every millimetre would result in too many lines. The scale would be cluttered and the individual lines would become confused. However, the situation is not hopeless, and it is possible to design a scale capable of providing this extra place of accuracy. Indeed, the most common solution is known as a Vernier scale.

Vernier scales are ingenious and simple and they are now present on almost all analogue devices from callipers, which measure length, to surveying instruments, which measure angles, to planimeters, and so on. They were invented by Pierre Vernier (1584–1638) and described in his book, *La Construction, l'Usage et les Propriétés du Quadrant Nouveau de Mathématique* ('The New Mathematical Quadrant') in Brussels in 1631. A quadrant is used to measure angles, and as the name implies it is a

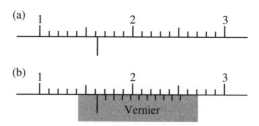

Figure 4.9. A Vernier scale.

quarter of a whole circle. As with so many other inventions that are discussed in this book, it languished almost unknown for many years after its original invention. For example, very few of the instruments in Bion's catalogue *Traité de la Construction et des Principeaux Usages des Instruments de Mathématique* (1709) (translated into English as Stone (1753)) contain a Vernier scale. We shall return to these in more detail in chapter 5.

Here, we shall describe a linear Vernier scale, rather than the original angular device, but the principles are the same. Imagine an equally spaced scale together with a pointer which provides a reading on an instrument. In our case this is a reading of length (see figure 4.9(a)). The pointer is just to the right of 1.6, so let us call this reading $1.6 + x$. Since the pointer does not line up with one of the scale markings, it is necessary to estimate the value of x. The Vernier scale is a method which reduces the amount of guesswork in this estimation, and consists of another scale which supplements this pointer. However, the Vernier scale is spaced at intervals of 0.09, not 0.1 as is the original scale in figure 4.9(a): see figure 4.9(b). To use this, count along to find the line on the Vernier scale which matches up most closely with a line on the top scale. In the figure above, this is the second line along the Vernier scale. Now, since the Vernier scale is spaced at intervals of 0.09 and the top scale is at 0.1 we have to solve the equation

$$x + 2 \times 0.09 = 2 \times 0.1,$$

obviously giving $x = 0.02$. Hence, the pointer is at position 1.62 on the top scale. In practice, you just count along the Vernier until the two scales coincide. Then you *know* the extra decimal

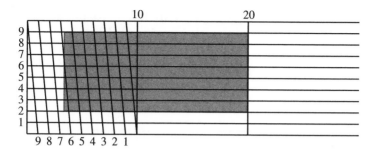

Figure 4.10. Obtaining extra accuracy with diagonal lines.

place, without having to estimate or perform the above calculation. Hence, the Vernier scale is an ingenious, simple and accurate mechanism.

There is another common mechanism which is often found
on rulers that can be used to gain an extra place of accuracy.
For the purposes of illustration, we shall assume that we have a
ruler with equal graduations one unit apart. Upon this draw ten
lines parallel with the edge of the ruler and one unit apart, with
one unit separating them. Next draw diagonal lines as shown
in figure 4.10. These diagonals intersect adjacent parallels one-
tenth of a unit apart. Of course, these diagonals need only be
drawn in one 10×10 unit square to use the whole length of the
ruler.

Imagine you need to use this device to measure the length of
something, such as the grey rectangle shown in figure 4.10. First
estimate the length to the nearest 10 units or so, in our case
between 10 and 20 units. Rather than placing one end of the
object to be measured at the zero of the scale as one usually
does, place it so that one end is on the 20 unit line and the other
end falls in the diagonally separated grid. Move the object up
and down until it lines up perfectly with a diagonal. This will
now allow a one-tenth of a unit length measurement to be read
off. In our case, the grey rectangle is 16.7 units long. In practice
these scales are rather hard to read. They are also very difficult to make and rely on being able to draw, originally by hand
and eye, very accurate diagonals. For a more accurate measurement of length, callipers with a Vernier scale are much better,

although correspondingly more expensive to produce, than a printed scale. Cutting along the horizontal lines for 0 and 10 and rolling the paper round so that these lines coincide gives us a screw thread. Such a screw forms the basis of micrometers and the fine adjustment mechanisms used in instruments of all kinds.

4.6   Similar Triangles and the Sector

Another important topic is that of similar triangles: two triangles are similar when the sets of angles are equal. When this happens, the lengths of corresponding sides are related by a fixed multiplication factor. Hence, similarity really has a lot to say about *scale*.

One classical mathematical instrument which is based on the properties of similar triangles is the sector, which is an instrument consisting of two rulers of equal length that are joined by a hinge. A number of scales are drawn on the instrument facilitating various mathematical and trigonometrical calculations—an example is shown in figure 4.11. This sector, approximately 12 in long, is made from ivory and brass. In common with many devices in this book, the issue of who first invented the sector is not without controversy. Its invention is often attributed to Galileo Galilei (1564–1642) in approximately 1597 (see, for example, Meskens 1997; Smith et al. 1996). However, Hopp (1998) claims that the sector was invented by a Guidobaldo de Monte, who was a friend of Galileo, as early as 1568. It turns out that the first scientific instrument to contain a logarithmic scale was the sector, not the slide rule. This was an innovation of Edmund Gunter, and its history is explained in section 11.2.

The sector can be used to perform many different numerical and plane or spherical trigonometrical calculations. The remainder of this section is a basic introduction to some of the calculations that are possible using a sector and an explanation of the most commonly appearing scales. There are two principal modes of operation: fully open as a straight rule, or partially open with the scales radiating out from the centre. At all times the sector is used in conjunction with a pair of compasses or dividers.

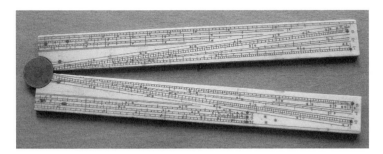

Figure 4.11. A sector.

When open fully, the long outside edge is marked with a scale divided into ten equal parts, each of these are further subdivided into ten parts. This scale is known as the *scale of decimals*. On one flat side close to the outside edge of the sector is a *scale of inches*. Typically a small pocket sector will be 12 in long when opened out. (If you ever have access to an old ivory sector, you might check to see how much shrinkage has occurred over the years.) On the other side are probably three long scales parallel to the long outside edge of the sector. These are marked N, S and T. The line N is a 'Gunter line', that is to say, a line marked in a logarithmic scale. The other two lines, S and T, are the lines of sines and tangents, respectively. These three scales are usually referred to as the *lines of artificial numbers*. The Gunter line, or logarithmic scale, is dealt with in chapter 11.

The other scales radiate out from the point where the two rules are hinged and are used with the sector open at an angle. Many of these scales are inscribed twice, once on each leg of the sector, and are referred to as *double scales*. These will be described individually in more detail below. When the compasses are used on a scale to measure a length from the centre it is called a *lateral distance*: distance x in figure 4.12, for example. When a measure is taken from any point on the line to its corresponding point on the line of the same denomination on the other leg, it is called a *transverse distance*: the distance y in figure 4.12, for example.

The sector derives its name from Euclid's *Elements* (book VI, proposition 4), which states that similar triangles have their like sides in proportion. To see why this theorem forms the mathematical basis of the sector, consider the two similar triangles

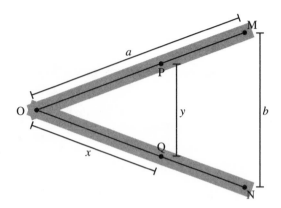

Figure 4.12. The principle of a sector.

shown in figure 4.12. We assume that the large outer triangle OMN has a small similar triangle OPQ within it. The line PQ is parallel to MN. The proposition can be used to deduce that $a/b = x/y$. The sector uses exactly this principle when the arms are open at an angle.

Multiplication can also be performed with the arms of the sector open. Locate the scales marked L which radiate along the arms from the centre and are marked in ten equal parts. Numbers can be multiplied using these scales by recalling figure 4.12, where the lines OM and ON represent the scales marked L on the legs of the sector and O is the point at which the legs pivot. Recall that $a/b = x/y$ or $ay = bx$. The length OM is always fixed at 10 units. We are left with three unknowns, x, y and b. Take a pair of compasses open to a distance which represent a length b. Using the compasses as a guide, open the sector so that the distance between M and N is the length b. Next locate the points P and Q that are a lateral distance x along the scale. Using the compasses take the transverse length y. This length can be read by placing one point at O and the other point on the line OM and reading directly on the evenly spaced scale. The length y that has been constructed is the answer to $10y = bx$. Division can be performed by an obvious reverse process.

For multiplying numbers this process is slow, requires a high degree of manual dexterity and is not very accurate. However, it is much quicker than the geometric construction given in

section 4.3. So, for a quick approximate and geometric calculation it does have limited practical use. Perhaps the main use
would be during a plane figure construction where it is necessary
to construct a line of a given proportion from one already on
the figure. This clearly removes the need to measure lengths
and multiply numbers. This shows that multiplication of lengths
can take place in an entirely graphical way. To understand the
difficulties it is really worthwhile constructing a simple sector, or
drawing a sector at a fixed angle and marking on the two scales.

Measuring Plane Angles

The sector can also be used to construct angles using the *lines
of chords*. This is a double scale, marked C, with scales that
run from 0 to 60. To protract an angle of $0 < t < 60°$ use the
scale C and open the sector so that the transverse distance MN
equals the lateral distance OM = ON. It is clear that the angle
MON = 60°. Using the compasses on the scale measure the lateral distance from O to t. Draw a circular arc, centred at O, from
M to N and place one point of the compass at N. Where the other
point intersects the arc will be marked R. The angle RON is t.
Angles can be measured using the reverse procedure. An angle
s greater than 60° can be protracted by repeatedly subtracting
60° or 90° from s.

Inscribing Polygons Inside a Given Circle

The typical sector also came with scales that could be used to
easily inscribe a regular polygon inside a circle of any given
radius using the *line of polygons*. This double scale, marked POL,
usually on the inside of the sector, is marked from 12 nearest
O down in unit steps to 4 at N and M, and these numbers refer
to the number of sides on the polygon itself. Hence, the sector could be used to draw polygons with seven, nine or eleven
sides, which the classical theory shows is impossible using only
a straight edge and compass.

Refer again to figure 4.12, this time assuming that ON represents the scale POL. Draw the circle in which you wish to inscribe

the polygon and keep the compasses open at the radius. Remember that the length of the sides of an inscribed hexagon is the radius of a circle. So, use the compasses to open the sector so that the transverse distance between the 6s on the POL scale equals the radius of the circle you have just drawn. The transverse distance between the figures is the lengths of the sides for an inscribed polygon with that number of sides. For example, to inscribe a square measure the transverse distance between the 4s on the POL scale and use this to walk the compasses round the circle. Then inscribe the polygon directly using these points.

It is also straightforward to construct a polygon with a given number of sides and a given side length, as opposed to an inscribed polygon. To do this, measure the length of the sides, y, with the compasses. Let P and Q be points on the POL scale corresponding to the number of sides of the proposed polygon. Open the sector so that P and Q are a distance y apart. The radius of the required circle is the transverse distance between the 6s. This is simply the reverse process although here the sector may not open wide enough.

Other Scales

The above gives some examples of the range and types of calculations that can be performed by the sector. The other scales can be used for a variety of other calculations, including spherical trigonometry, that are not discussed in detail here. A full description of how to use such a sector can be found in, for example, Heather (1871).

The sector was a standard component in a set of mathematical instruments. However, the smaller pocket sectors would have been of little or no use because of the difficulty in using them accurately. Compass points damage scales with regular use and this reduces the accuracy of the instrument. The large number of well-preserved sectors that exist would add weight to the school of thought that claims they were of little practical use, other than for school geometry problems. However, large instruments were used by artillery officers for calculations in the field.

Chapter 5

DIVIDING THE CIRCLE

Devyde it into sixty parties equals.

Chaucer (1872)

Length is a measure of the extent of a one-dimensional line. In defining length in chapter 4 we had considerable freedom in taking an arbitrary unit length. It was then a geometric task to divide this into fractions and to build the world's first ruler.

In this chapter we measure the extent of *rotation*, and here we have a natural base unit of one complete rotation. The choice now is how to divide the circle. The quotation from Geoffrey Chaucer (1343–1400) above, from the first scientific book published in England in the vernacular rather than in Latin, describes dividing angles on the astrolabe. An astrolabe is a device with which astronomical measurements can be taken, and it is no coincidence that astronomers were, as a profession, most interested in accurate angular measurement. Of course, navigation is closely related to astronomy and hence finding one's position using a sextant also relies on accurate measurement of angles. Surveying and cartography also depend on being able to *triangulate* one's position. Keay (2000) gives a very good idea of the care and precautions taken to set up the datum line $7\frac{1}{2}$ miles (12 km) long as the basis for the meridian survey of India in the nineteenth century. From the measured angles and a datum line of known length the mathematical task is to reconstruct the ground surveyed. All this relies on accurate measurements.

All of these applications are only possible if accurate instruments are available with which to undertake the measurements. Progress in science goes hand in hand with progress in the development of accurate and reliable instruments, although in the history of science the importance of instrumentation is not always given its due weight. As Chapman (1990) succinctly puts it:

> Instruments came to be seen not only as refinements or aids to human perception, but as the very arbiters in scientific discussion, when the acceptance or otherwise of a theory depended upon the interpretation of observational evidence.

The invention of the telescope by Galileo Galilei opened up new worlds to science. The use of the telescope as an *observational* tool is usually given prominence, for example to view the moons of Jupiter. Certainly less immediately glamorous are its uses for *measurement*, but it is this which really drove science forward.

Before the invention of the telescope, astronomers such as Tycho Brahe (1546–1601) relied on naked eye measurements. Essentially there are two problems in astronomy. The first is in observing the object in the sky, and lining up an instrument upon it. The second is in recording the position of the instrument, for example by reading the angular position. Originally it was sufficient to use cross-hairs, since the accuracy of the whole process was determined by the accuracy of the angular scales. Gradually the design of such scales was improved so that greater accuracy could be achieved. For example, the system of diagonal lines illustrated in section 4.5 was applied to instruments such as the octant shown in figure 5.1. Such improvements in scale design meant that the astronomer's overall accuracy was hampered by his ability to see an object in the sky and line up the measuring device to take a reading.

Experiments by Robert Hooke (1635–1703), repeated in more recent studies, showed that the naked eye could resolve an object that occupies about $1'$ (one minute) of arc. Kepler, for example, was concerned with a difference of $8'$ between the position of a planet predicted by a circular orbit and what was actually observed. It was only with Tycho Brahe, the last and greatest naked eye astronomer, that orders of accuracy approaching $1'$ *total error* were achieved. Hence, the level of accuracy that the

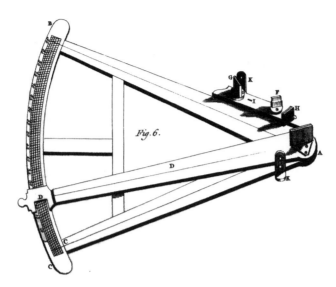

Figure 5.1. An eighteenth-century octant from Stone (1753).

naked eye can achieve is simply not sufficient for testing astronomical theory.

It is the Yorkshire mathematician William Gascoigne (1612–44) who is credited with putting cross-hairs into a telescope to create a telescopic sight. The key to this is to place the cross-hairs inside the telescope at the instrument's common focus. Gascoigne was observing the Sun when he noticed that a spider had spun a web inside the telescope. By chance this was in sharp focus, and from this discovery Gascoigne developed a practical telescopic sight. The practical use of this was obvious to him and his friends, and it is reported in Derham (1717, p. 604) that when the astronomer John Horrox first saw the invention it 'ravished his mind quite from itself and left him in an extasie between admiration and amazement'.

Lenses were also applied to read scales more accurately, and the astronomer Ole Römer (1644–1710) used a low-power microscope containing a number of hairlines to remove the problems of two scales wearing against each other. Gascoigne also developed a device consisting of two movable cross-hairs, controlled by a graduated screw. By measuring the number of turns, the distance between the two cross-hairs could be calculated and from this the angle between two objects which were very close together could be calculated. This was used by him to measure the angular width of the Moon and the planets. Of course, the accuracy of this device depends crucially on the accuracy of the threads of the screws, which in the early seventeenth century were extremely difficult and expensive to make by hand. Screw threads were also used to move and position telescopic sights accurately, although early designs by Robert Hooke and others were not particularly successful from a practical point of view. The Astronomer Royal John Flamsteed (1646–1719) complained bitterly that he was 'much troubled with Mr Hooke who, not being troubled with the use of any instrument, will needs force his ill-contrived devices on us' (Flamsteed 1725, p. 103).

5.1   Units of Angular Measurement

It is well worth pausing for a moment to review mathematical systems for angular measurement. The most commonly used system for measuring angles is the division of a circle into three hundred and sixty degrees, i.e. 360°. Each degree is subsequently divided into sixty *minutes* and each minute into sixty *seconds*. An angle is then written as $17°\ 34'\ 12''$. Each minute is $\frac{1}{60} \approx 0.0167$ of a degree, and each second is $\frac{1}{3600} \approx 0.000\,278$ of a degree. As we have seen, dividing a circle into six equal parts is simple, and can be done accurately. Then subdividing each into sixty equal parts to give degrees and then again into minutes and seconds is a remnant of the ancient Babylonian base-sixty number system. This division into 360° is also convenient from a practical point of view. For example, on a basic school protractor with a diameter of 90 mm a single degree occupies an arc length of about 0.75 mm on the edge. This is perfectly easy to distinguish by eye, and yet still small enough a unit to be practically useful. Since 60 and 360 have so many integer factors, many fractions of a rotation are easy to express exactly as whole numbers of degrees.

Mathematicians use a different system for measuring angles and it is known as the *radian*. In this system a circle is divided into 2π radians. The advantage of this is that many formulae become extremely simple and elegant to express. For example, in a circle of radius r take an arc with angle t, measured in radians. The arc length is simply rt. The corresponding formula, with t in degrees, is $\frac{1}{180}\pi rt$. However, since π is an irrational number there is no longer a whole number of radians in a whole number of full rotations. From a practical point of view this is a catastrophe.

Another unit for measuring angles is the decimal degree. In this scheme a right angle is divided into 100 units, and many modern scientific calculators implement these units as *grads*. For military purposes the artillery divide a full circle into 6400 parts, called *mils*, not to be confused with the spoken abbreviation for a millimetre.

It is clear that astronomical tables of angles were not used solely by their original observer but were intended for publication for the benefit of other astronomers and navigators. In these circumstances it is vital that there should be a widely understood and accepted unit of angle, and we naturally base our protractor on a circle of 360°.

We shall see in section 5.4 that, from the practical point of view when trying to engrave a graduated scale using only geometric constructions, 90° in a quarter of a rotation is not as convenient as the number of factors would first suggest. In particular, it is always possible to divide any angle exactly into two equal parts, but it is in general impossible to divide an arbitrary angle into three. Furthermore, we know that 60° can easily be marked to a high degree of accuracy, using the radius of the circle. From this, 30° and 15° can be found by simply bisecting repeatedly. To go any further we need to divide by three and then five, which poses a problem.

Instead, George Graham (1675–1752) chose to divide a right angle into 96 equal units. In this, the radius will strike off 64 units of angle, and since this is a power of two the remainder can be repeatedly halved to get single angular units. He used this scheme on an 8 foot quadrant in the Royal Observatory in Greenwich. This was engraved with both a 90° scale and an adjacent scale of 96 units. These two scales allowed a cross-check, and the

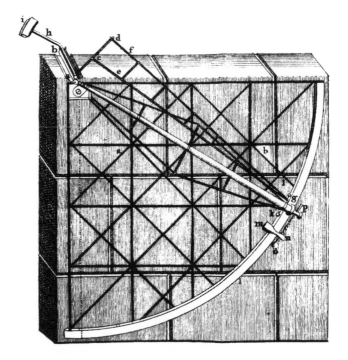

Figure 5.2. A mural quadrant from the
Royal Observatory in Greenwich.

whole construction was carried out with such skill that the two
arcs differ by no more than 6″, which is approximately 0.0017°.
The sketch of this quadrant given in Stone (1753) is reproduced
in figure 5.2.

While the 96 point scale is an elegant solution to the problem of geometrical construction, it is somewhat unsatisfactory.
Users want accurate scales based upon the familiar 90°. It is
awkward, and a potential source of error, to convert between
the two systems. It was John Bird (1709–76) who solved this
problem, and he was also one of the first practitioners to leave a
detailed account of his own methods of working: Bird (1767),
which was written for the Board of Longitude. He developed
a system based upon 85° 20′ and to see why this works we
note that $85 \times 60 + 20 = 2^{10} \times 5$. Hence, by repeatedly dividing the angle 85° 20′ in half we eventually arrive at an angle
of 5′. It only remains to construct the crucial angle of 85° 20′

sufficiently accurately. He did this by accurately constructing $30°$, $15°$, $10°$ $20'$ and $4°$ and adding them together. In particular, he drew a large circle and calculated and measured the lengths of the chords subtending some of these angles, measuring them to $\frac{1}{1000}$ th of an inch. This introduced the possibility of error, and it is a testament to his professional skill that such a method could actually be made to work. A large number of independent cross-checks were then undertaken to gauge the accuracy of this one angle, before the division of this angle into halves to create a scale of degrees. According to Ludlam (1786), this technique was a great success, both in the Greenwich observatory and in Bird's many other instruments.

When working at this level of accuracy the temperature of the workplace matters. If the metal being engraved and the compasses are made of different materials, then not only must they be at the same temperature but also at a well-defined temperature. Otherwise, problems will arise with differential expansion. In metrology, $20\,°C$ is now taken to be a standard working temperature.

After the 1780s astronomy moved away from large fixed quadrants to the use of full graduated circles. Whereas the division of the quadrant is essentially a geometrical exercise, dividing the circle is mechanical, and so we do not pursue the practical problems of astronomy further. Rather, we consider the geometry and eventually show how to trisect an angle by other means, something which practitioners such as Bird had assiduously avoided doing.

While this chapter concentrates on the mathematical problem of dividing the circle, the preceding discussion gives some background to the associated practical problems with constructing an astronomical instrument, which is very much more than an accurate protractor.

5.2 Constructing Base Angles via Polygons

It would be tempting for our application to mark out the smallest angle to a supremely high degree of accuracy and use this to mark out all others by walking the protractor around the circle.

This would result in a accumulation of errors. Instead we work from the other direction or dividing larger angles into parts. We first note that there is a strong connection between angles and regular polygons. We notice immediately that if we can construct an n-sided regular polygon, we can construct an angle of $360°/n$. Conversely, if we can (somehow) construct an angle of $360°/n$, then we can construct an n-sided regular polygon inside a circle. So, at least to begin with, we concentrate on building regular polygons.

It is straightforward to construct shapes such as equilateral triangles, squares and regular hexagons. For the hexagon, for example, start by drawing a circle and mark some starting point A_0 on this (see figure 5.3). Keeping the compass radius fixed, draw a second circle centred at A_0. Where this intersects the original circle mark A_1 and A_5. Next draw a circle centred at A_1. This will intersect the original circle at A_0, but also at a new point A_2. Repeat the process centring a circle at A_2 to generate A_3. This is similar to walking around the circle with the compasses to mark out successive points. In fact, one does not need to draw the entire circle at each stage and only small arcs have been included in figure 5.3 to keep the diagram uncluttered. Walking around a circle with a pair of compasses in this way is exactly how we will construct regular polygons in general. We only need to construct the length of one edge and the circle in which the polygon is to be inscribed.

Let us refer to an n-sided regular polygon simply as an n-gon for brevity. Assuming we can construct an n-gon, then we can certainly construct a $2n$-gon, a $4n$-gon, etc., by halving angles. Similarly, if n is not prime, say $n = pq$ where p and q are integers, we can also easily construct a p-gon by simply skipping every q vertices. For example, to construct a 3-gon (better known as an equilateral triangle) from a regular hexagon, miss out every other vertex. From this observation we see that it will be important to consider the 'prime-gons'. The following theorem shows exactly which n-gons are constructible using only a straight edge and compass, but the proof is not included.

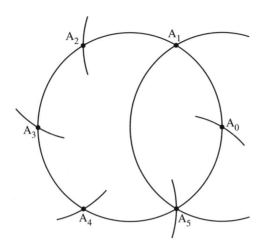

Figure 5.3. Constructing a regular hexagon.

Theorem 5.1. *An n-gon is constructible using straight edge and compass if and only if*

$$n = 2^k f_1 \times \cdots \times f_m,$$

where the f_j are different Fermat prime numbers. That is to say, f_j are prime numbers of the form $2^{2^j} + 1$ for some $j = 0, 1, \ldots$.

To properly understand how restrictive this theorem is, let us consider which numbers of the form $2^{2^j} + 1$ are prime. The sequence is

j	$2^{2^j} + 1$
0	3
1	5
2	17
3	257
4	65 537
5	4 294 967 297

This sequence looks promising, since 3, 5, 17, 257 and 65 537 are all prime. However, $4\,294\,967\,297 = 641 \times 6\,700\,417$, so it is not prime. Indeed, so far no other primes of the form $2^{2^j} + 1$ are known, and whether there are any or not is an open problem. So,

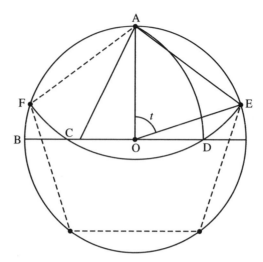

Figure 5.4. Constructing a regular pentagon.

currently there are only five choices for the f_j in theorem 5.1, although we can take any combination of these or none at all. This is really quite limiting because by telling us which n-gons are constructible, the theorem tells us which basic angles, or rather fractions of a circle, are not constructible.

Hence, we have a very limited range of base angles from which to build our protractor. From a practical point of view, adding angles is not a sensible strategy, since errors are cumulative. More importantly, there is no way of obtaining an *independent check* of the scale. Hence, for our application the most important task is to divide angles. Once all angles have been marked out they can be checked against each other.

5.3 Constructing a Regular Pentagon

To illustrate this excursion into polygons we shall explain the construction of a regular pentagon, shown in figure 5.4. Let us take a quadrant AOB in a unit circle, i.e. OA = 1 and the angle AOB is 90°. Mark C as the midpoint of the radius, then by the Pythagorean theorem

$$\text{AC} = \sqrt{1^2 + (\tfrac{1}{2})^2} = \tfrac{\sqrt{5}}{2}.$$

With C as the centre draw an arc of radius AC to cut the diameter OB at D, so that

$$OD = \frac{\sqrt{5} - 1}{2}.$$

As a complete digression you will notice that OD/OB is the Golden Ratio. Using the Pythagorean theorem again,

$$AD = \sqrt{1^2 + \left(\frac{\sqrt{5} - 1}{2}\right)^2}.$$

The final step is to draw an arc of this radius from A cutting the circle at E and F. Triangle OEA is isosceles and so we can write

$$\sin(\tfrac{1}{2}t) = \frac{AE/2}{1},$$

giving $t = 72°$. AE and AF are therefore the sides of a regular pentagon.

Although it is not a Euclidean construction, we should mention that one very neat alternative way to construct a regular pentagon for yourself is to take a strip of paper with parallel sides and tie a normal overhand knot. Carefully tightening and flattening this knot leaves a regular pentagon. Constructing a seventeen-sided shape is actually quite a complicated straight edge and compass construction, and one shudders to imagine trying to construct the 257- and 65 537-sided figures in practice. Indeed, even though a polygon of 65 537 sides is of no help to us in constructing a protractor, do not be tempted to try it out of general interest.

A too-persistent research student drove his supervisor to say 'Go away and work out the construction for a regular polygon of 65 537 (= 2^{16} + 1) sides'. The student returned 20 years later with a construction (deposited in the archives at Göttingen).

Littlewood (1986)

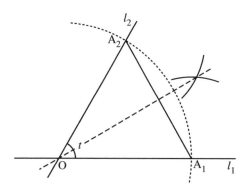

Figure 5.5. Dividing the angle t in half.

5.4 Building the World's First Protractor

The object of this section is to discover how far it is possible to graduate a protractor, such as a typical school protractor, into units of one degree using only exact Euclidean constructions involving a straight edge and compass.

The symmetry of a circle is such that we can confine our description to constructing a quadrant, and this can easily be constructed by starting from a straight line: an angle of 180°. Taking a vertex wherever we please on the line, the next constructional step is to bisect this angle to obtain an angle of 90°. This can be done by constructing a perpendicular to the line, just as in figure 4.2.

We have claimed that it is possible to divide an arbitrary angle into two equal parts. To show how, we begin with an angle t between two previously constructed lines l_1 and l_2, which meet at O. Such a configuration is shown in figure 5.5. To divide this angle in half draw a circle centred at O and where this meets l_1 and l_2 mark A_1 and A_2. Connect these with a straight line, l_3. As detailed above, construct the perpendicular to l_3 through O, shown in figure 5.5 as the dashed line. That this line divides the angle t in half is easily proved using similar triangles. This procedure is known as bisecting an angle.

The prime factors of 90 are 2, 3, 3 and 5, so we can bisect 90° to reach 45°. No further bisection is possible because we do not wish to introduce fractions of a degree into our quadrant.

This leaves us with the factors 3, 3 and 5 and it is impossible to trisect an arbitrary angle using only Euclidean tools, so we must find some alternative method. Since we are interested only in exact trisection, which is generally impossible, we defer discussion of approximate and mechanical methods to a later section.

We can, though, construct an angle of 60° by the familiar technique used to define the vertices of an equilateral triangle and so we have effectively trisected 90°, proving that it is possible to trisect some angles. 60° has two factors of 2 so that we end up with graduations at 15° intervals. This is an opportunity to check the precision of our work: 45° was first constructed by bisecting 90° and we have now produced the same angle independently by twice bisecting 60°. We are not making a large astronomical quadrant and so we do not really need this check, but on any precision instrument any way of verifying our work as it progresses is very welcome.

We can use part of the regular pentagon, giving angles of 72° and 18° to add to our quadrant. 72 has prime factors 2, 2, 2, 3 and 3 so we can now add angles at 9° intervals. Furthermore, since the difference between 72° and 60° is 12°, we find that some 3° graduations are possible after further bisections.

In principle, but maybe not in practice, we could repeat the constructions based on 60° and 72° on two of the angles we have already marked on the quadrant. The result, after some bisections, is a full scale of divisions 3° apart. Whatever we do though we cannot further divide 3° into units of 1° so we must resort to approximate methods.

There is, however, another set of practical problems associated with the width of the lines. Harking back to chapter 1, in all instruments we are relying on the intersection of two lines of finite width to define points through which a radial line must pass, or to form centres for further constructions. To put this into an everyday context, the interval between 1° graduations is about 0.75 mm on the circumference of a typical school protractor. A commonly used drawing instrument is a pencil with a 0.5 mm diameter lead.

As a general rule for locating a point, it is always best if the two intersecting lines cross at right angles, as its centre is most sharply defined when the overlap is square rather than an elongated parallelogram. Another point to remember is that if we draw our protractor or quadrant with a pencil on paper we can always ink in the lines we wish to keep and rub out the others. This is a luxury that is not so readily available if the working material is plastic or metal. One final general remark: draw your quadrant as large as possible. As error in circumferential position of 0.75 mm is equivalent to 1° on a school protractor but on an 8 foot radius quadrant it is about 1′ of arc, a standard well surpassed by the original craftsmen.

5.5 *Approximately* Trisecting an Angle

Can we divide an arbitrary angle into three equal parts? A naive first attempt would be to adapt the construction in figure 5.5. That is to say, we take the arbitrary angle we wish to trisect and draw the chord and divide this into three equal parts. One might well expect this construction to trisect the angle. To show that this is not the case, consider what happens when we apply this construction to a right angle.

Here, we simply draw an arc of radius 1, across the right angle, and then draw a chord. This is divided into three equal sections. We shall show that the angle marked t in figure 5.6 is *not* 30°, proving that this method does not trisect the angle.

Using the cosine formula, we can calculate that $l = \frac{\sqrt{5}}{3}$, and from this the sine rule gives us that $\sin(t) = \frac{1}{\sqrt{5}}$. This gives $t \approx 26.6°$, and not 30°, giving quite a large error from the true positions, which are shown by the dots on the circular arc. To examine this error in more detail, assume we have the angle AOB shown in figure 5.7 as s.

For simplicity, we assume that OA = OB = 1 and the procedure as before is to join AB with a straight line. Next divide this line into three equal sections, and mark points P_1 and P_2 on the line. We shall determine the angle t, and in particular consider how close t is to $\frac{1}{3}s$.

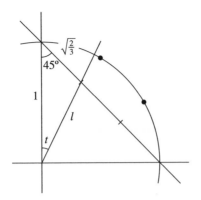

Figure 5.6. Trisecting the chord of a right angle.

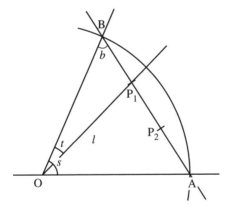

Figure 5.7. Attempting to divide an
arbitrary angle into three: *trisection.*

The first thing to note is that (i) the angle b is equal to $\frac{1}{2}(\pi - s)$
and (ii) the length

$$BP_1 = \tfrac{2}{3} \sin(\tfrac{1}{2}s).$$

From this and the cosine rule we can establish that

$$l^2 = 1 - \tfrac{8}{9} \sin(\tfrac{1}{2}s)^2,$$

and in turn from the sine rule that

$$\sin(t) = \frac{2}{3} \frac{\cos(\tfrac{1}{2}s)\sin(\tfrac{1}{2}s)}{\sqrt{1 - \tfrac{8}{9}\sin(\tfrac{1}{2}s)^2}}.$$

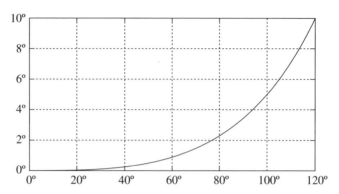

Figure 5.8. The error $\frac{1}{3}s - t$ against angle t when approximately trisecting an angle.

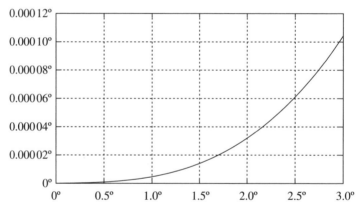

Figure 5.9. The error $\frac{1}{3}s - t$ against angle t when approximately trisecting a small angle.

This formula gives the exact relationship between s and t. For values of s between 0° and 120°, the graph of the error, that is the difference between $\frac{1}{3}s$ and t, is shown in figure 5.8. Notice that for s between 0° and 60°, the actual error is less than 0.9°, but the central angle of the triplet has an error of about 1.8°. This is an accuracy which most people would be hard pushed to achieve using a normal protractor, so that for practical purposes with small angles this method will probably suffice; indeed figure 5.9 shows the error for values of s between 0° and 3°. A word of caution though. Trying to divide a short line exactly into three is also difficult. Here, proportional dividers can help. Indeed,

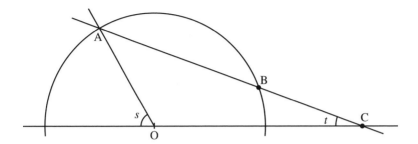

Figure 5.10. Details of angle trisection with a marked ruler.

pricking the surface with the points of such dividers can provide more accurate locations for the points than a pencil line would provide.

This initial method does not work exactly, and although we do not prove it here, using only Euclidean constructions it is impossible to trisect an arbitrary angle exactly. Despite this unfortunate fact, there are even better approximate methods using only a ruler and compass. Some give extremely good approximations that are certainly better than any protractor. The method of Fahmi (1965) is a typical example.

5.6   Trisecting an Angle by Other Means

There are other ways to exactly trisect an arbitrary angle, and in this section we begin by showing how specialized the setting of ruler and compass really is. We do this by adding one extra tool. This is simply a straight edge marked with two points a distance r apart. We could take r to be 1, but in fact this does not matter— any two marks on the ruler will suffice. So, this modified straight edge is really the most basic of all rulers. Using this extra tool we explain how to trisect an arbitrary angle s.

We start by drawing a baseline and another line between which lies the angle s (see figure 5.10). On this we draw a circle, centred at the intersection of the lines and with a radius r. Place the marked ruler with the edge running through the point A and so that the distance BC is the marked distance r. We will examine the angle at t.

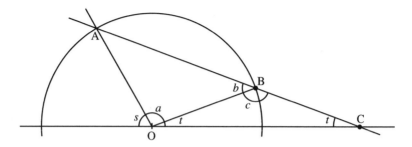

Figure 5.11. Trisecting an arbitrary angle s.

To determine t, we draw a further radius to where the line AC cuts the circle at B and mark the angles a, b and c as shown in figure 5.11. Since the two triangles AOB and OBC are isosceles, we have that

$$2t + c = b + c,$$

so $b = 2t$. Next

$$2b + a = s + a + t,$$

so $3t = s$.

This is known as Pascal's angle trisector, and we shall describe a simple mechanism that incorporates it in the next section.

5.7   Trisection of an Arbitrary Angle

Fortunately, from a practical point of view, you can forget about protractors and the impossibility of trisecting an angle by Euclidean constructions. In this section we shall show a way to do it and much more besides by means of a simple linkage. The linkage can be made without the need for measurement, its actual size being dictated solely by the available materials. It consists of two straight channels or guides of ⊔ cross-section hinged together, some pivots that can slide in these guides freely without wobbling, and a set of links. The critical feature is that the centre of the linkage pin must lie on the centre line of the guides. The links should also all be the same length.

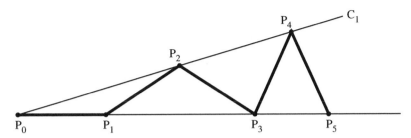

Figure 5.12. Dividing an angle by five.

Figure 5.13. Generalizing the trisection construction.

Only simple geometrical knowledge is needed to understand how it works. First recall that the base angles of an isosceles triangle are equal. Second we note that the extension angle of a triangle equals the sum of the two opposite internal angles. We shall adopt the convention that links are shown as broad lines and the centre lines of the guides as thin lines. The linkage pin acts as a pivot, marked P_0, and is used in all constructions.

Three sliding pivots are needed, together with three links. We can see from plate 12 how this device physically implements Pascal's angle trisector. We can expand on this by adding more pivots and links, an example of which is shown in figure 5.13. In each case the final angle can be divided into equal parts.

Figure 5.12 shows that the angle

$$P_2P_0P_1 = \tfrac{1}{5}C_1P_4P_5.$$

In the model in figure 5.13, the angle between the guides is not fixed, indeed it would be useless if it could divide only one

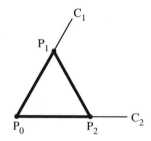

Figure 5.14. Constructing 60°.

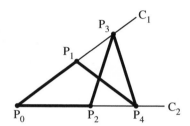

Figure 5.15. Constructing 36°.

particular angle between the guides. The numbers of pivots and links used is determined not by geometry but by their physical sizes. Unless they are small the maximum potential number of links is about nine or eleven.

A whole range of other geometrical games can be played if the linkages are put together in such a way that it becomes a rigid framework. In a rigid framework the angles are fixed. The game is to find a framework containing a specific angle, such as the construction of 60° shown in figure 5.14. The equilateral triangle made by the three links cannot be deformed and so the angle is set at 60°. All this is obvious, so we now add two more pivots to these links as shown in figure 5.15. As with the angle dividers we apply the properties of isosceles triangles to find

$$C_1 P_0 C_2 = 36°.$$

Noticing that $60° = \frac{1}{3}180°$ and that $36° = \frac{1}{5}180°$ points the way to generalizing this construction. Indeed, we do find that if we

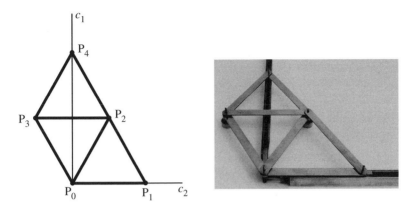

Figure 5.16. Constructing 90° with external supports.

continue the pattern in a natural way,

$$C_1P_0C_2 = \frac{180°}{2k + 1},$$

where $2k + 1$ equals the number of links and $2k$ is the number of pivots. In principle, then, we have a means of constructing all the odd unit factors of 180°, although the same limitation is imposed by the physical size of the links and pivot slides.

Setting the guides to an even unit fraction of 180° is a completely different story, and there is no general method. The best we can do is 30°, or $\frac{1}{6}180°$. If we make two sets of linkages, we can subtract angles, but this is not all that helpful. Instead we go back to one linkage, but this time we will make use of some more pivots that can lie *outside* the guides, and construct rigid frameworks for 90° and 45°. For example, one way to construct 90° is to use two external supports and seven linkages: see figure 5.16.

Accepting that these are valid constructions, we can look back at 45° and see that we have constructed an angle $\tan^{-1}(1)$, so we could also form $\tan^{-1}(\frac{1}{2})$, or, more generally, $\tan^{-1}(1/n)$, where n is the number of links lying along the bottom guide. There are many other possibilities here—you can amuse yourself by finding them.

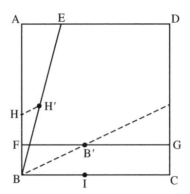

Figure 5.17. The origami trisection of EBC.

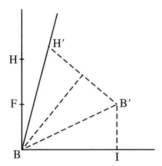

Figure 5.18. Proving that the construction trisects the angle.

5.8 Origami

Another way to trisect an arbitrary angle is using origami, something you could try out for yourself with an ordinary square of paper.

To trisect an angle take a square ABCD and on the edge AD mark a point E as shown in figure 5.17. Our task is to exactly divide the angle EBC into three. To construct this angle we proceed as follows. First choose a point F on the line AB and fold the line FG parallel to BC. Next, mark the point H where B meets the line and unfold the square. Now create a diagonal fold so that the point H lies on the line BE and the point B lies on the line FG. Where these points meet the lines mark H′ and B′, respectively. Again, unfold the square. Our claim, which we shall prove in a moment, is that the angle EBC is exactly three times the angle B'BC.

We first note that the distance BH is identical to B'H'. Furthermore, this distance is twice the distance FB = B'I. Next, note that the angle FBB' = BB'I, and furthermore this angle is also equal to BB'H'. If we drop a perpendicular to B'H' through B, we see that the three right-angled triangles shown in figure 5.18 are congruent. Hence the angle EBC is exactly three times the angle B'BC.

Chapter 6

FALLING APART

Humpty Dumpty sat on a wall.
Humpty Dumpty had a great fall.
All the King's horses and all the King's men
Couldn't put Humpty together again.

Nursery rhyme

In this chapter we discuss mathematical jigsaws based on plane dissections and other solid models. Not only are these fun, but they also illustrate some important basic mathematical results. Mathematical dissections such as these all rely on the concepts of area and volume. We are concerned here with simple geometric shapes but we shall address the more general problem of measuring the area of an irregular shape in chapter 8.

6.1 Adding Up Sequences of Integers

There is a story about the famous mathematician Carl Friedrich Gauss (1777-1855) that while at primary school he was able to add the numbers from 1 to 100 in his head, and simply wrote the answer on his slate. He did this by noticing the following pattern, which can be adapted to add together the first n integers. That is to say,

$$1 + 2 + 3 + 4 + \cdots + n.$$

The idea is simply to write the numbers twice, one list on top of the other but with the order reversed:

$$\underbrace{\begin{array}{ccccccccccc} 1 & + & 2 & + & 3 & + & \cdots & + & (n-1) & + & n \\ n & + & (n-1) & + & (n-2) & + & \cdots & + & 2 & + & 1 \end{array}}_{n \text{ terms}}.$$

Notice that the total in each column is $n + 1$, and there are n such columns. Thus

$$1 + 2 + 3 + 4 + \cdots + n = \tfrac{1}{2}n(n + 1). \tag{6.1}$$

This result may easily be seen by taking two sets of rectangular rods, consisting of one of each length from 1 to n, and pairing them up, so that the longest, i.e. that of length n, is paired with the shortest, i.e. that of length 1. (Sets of coloured rods, known as Cuisenaire rods, are standard equipment in many primary classrooms and can be easily employed for this purpose.) This is illustrated in plate 13. For somewhat obvious reasons the numbers $\tfrac{1}{2}n(n + 1)$ are sometimes referred to as *triangular numbers*.

Next we turn our attention to summing the cubes. That is to say adding

$$1^3 + 2^3 + 3^3 + \cdots + n^3.$$

It is possible to prove that

$$1^3 + 2^3 + 3^3 + \cdots + n^3 = (1 + 2 + 3 + \cdots + n)^2,$$

which is sometimes expressed by saying that 'the sum of the cubes of 1 to n is the same as the square of their sum'. This is known as Nicomachus's theorem, and is attributed to Nicomachus of Gerasa, who is believed to have proved it around 100 C.E. In fact, our first result can be used here to show that this equals $\tfrac{1}{4}n^2(n + 1)^2$. Figure 6.1 shows one particularly satisfying illustration of why this is true, in the case $n = 4$. Although some of the squares do overlap, the total area is $(1 + 2 + 3 + 4)^2$. If we count the individual squares we see that

$$1 \times 1^2 + 2 \times 2^2 + 3 \times 3^2 + 4 \times 4^2 = 1^3 + 2^3 + 3^3 + 4^3 = (1 + 2 + 3 + 4)^2.$$

We have missed out summing the squares of the integers. It is also possible to prove that

$$1^2 + 2^2 + 3^2 + \cdots + n^2 = \tfrac{1}{6}n(n + 1)(2n + 1). \tag{6.2}$$

Since this is a cubic in n, and contains the fraction $\tfrac{1}{6}$, one naturally considers whether it is possible to arrange six sets of flat

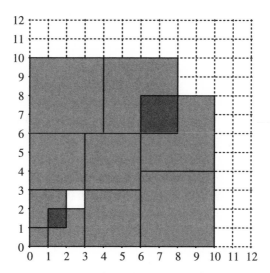

Figure 6.1. Summing the cubes of the first four integers.

squares into a box shape. Each set consists of one unit square, one 2×2 square, one 3×3 square, etc. And, indeed, if one has the patience to cut out the required number of unit cubes and stick them into squares, this can be done. For example, taking $n = 5$ we can find an arrangement of these squares into a $5 \times 6 \times 11$ block. Actually, an even more surprising and satisfying three-dimensional jigsaw can be made with six copies of an off-centre pyramid, in which the number of steps can be made arbitrarily large. This is illustrated in plate 14.

Anyone who makes six copies of this pyramid and puts them together will see that another hybrid three-dimensional jigsaw can be made from only three of these and a stepped triangular piece taken from half of the illustration that $1 + 2 + \cdots + n = \frac{1}{2}n(n + 1)$, shown in plate 13.

6.2 Duijvestijn's Dissection

Ideally we would like to make a flat square jigsaw consisting of smaller squares. If we want to take a complete sequence we need to know whether it is possible to cut a square up into a sequence of other squares of size $1^2,\ldots,n^2$. This will only be possible if

Models

Plate 1. Dudeney's dissection

Plate 2. Chebyshev's straight-line mechanism, in original and alternative forms.

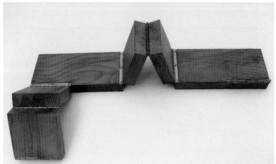

Plate 3. Sarrus's linkage.

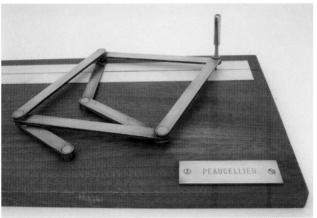

Plate 4. Two forms of Peaucellier's linkage.

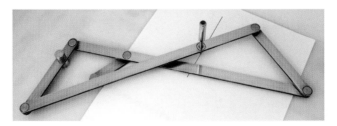

Plate 5. Hart's straight-line linkage.

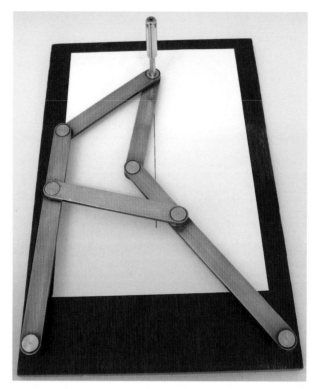

Plate 6. Hart's A-frame linkage.

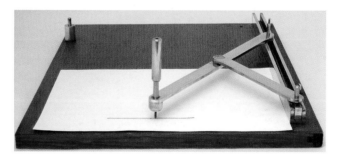

Plate 7. Scott Russell's linkage.

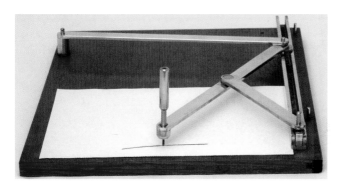

Plate 8. The Grasshopper linkage.

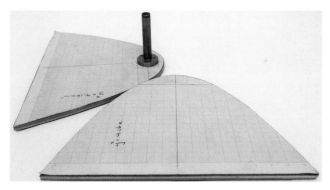

Plate 9. Rolling two parabolic templates to draw a straight line.

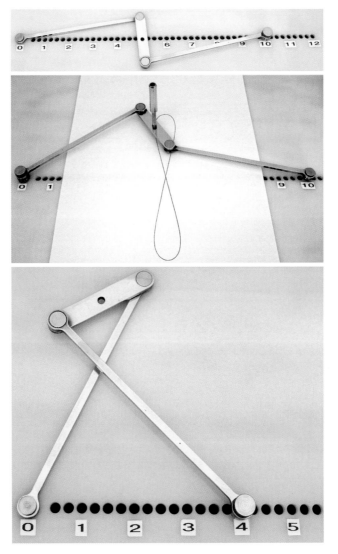

Plate 10. General four-bar linkages.

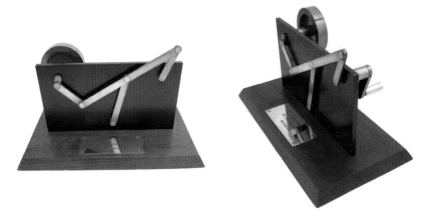

Plate 11. Chebyshev's six-bar paradoxical mechanism.

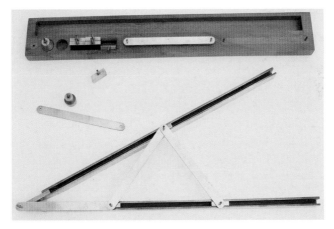

Plate 12. A model which includes Pascal's angle trisector.

Plate 13. Illustrating the sum from 1 to n.

Plate 14. Summing the squares of the numbers from 1 to n.

Plate 15. Duijvestijn's dissection.

Plate 16. A hemipseudosphere.

Plate 17. Three-dimensional shapes of constant width in wood, brass and plastics.

Plate 18. Three-dimensional shapes of constant width.

Plate 19. A three-dimensional shape of constant width based on a regular tetrahedron.

Plate 20. Watts's drill that cuts a square hole.

Plate 21. A drill that cuts a square hole.

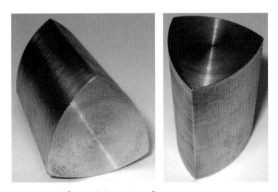

Plate 22. A Reuleaux rotor.

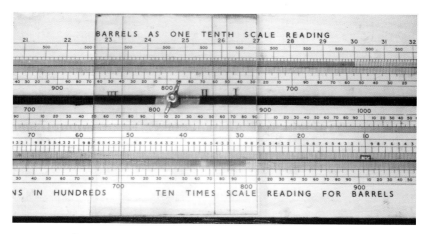

Plate 23. The Magnameta Oil Tonnage Calculator.

Plate 24. A balancing stack of wooden blocks,
90% to the point of collapse.

Plate 25. Balancing stacks of wooden blocks, 50%, 70% and 90% to the point of collapse.

Plate 26. Twisting part of the stack.

Plate 27. Self-righting (left) and over-balanced (right) stacks.

Plate 28. A 'two-tip' polyhedron.

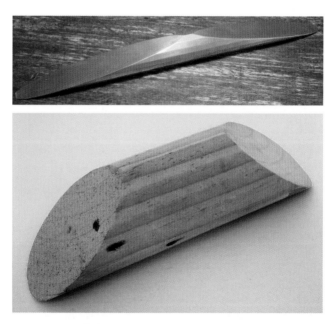

Plate 29. Uni-stable polyhedra.

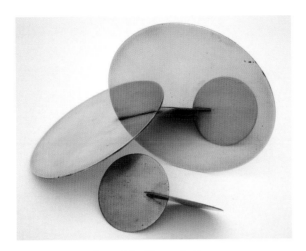

Plate 30. Slotted discs.

Plate 31. Alternatives to two slotted discs.

Plate 32. Slotted ellipses.

Plate 33. Two super-eggs.

the total area is a square number. That is to say, if

$$1^2 + \cdots + n^2 = \tfrac{1}{6}n(n + 1)(2n + 1) = r^2$$

for some integer r. In fact there are only two integers n for which this is possible: $n = 1$ and $n = 24$ (giving $r = 70$). That is to say, $1^2 + 2^2 + \cdots + 24^2 = 70^2$. However, simply because the shapes have the same area does not mean we can cut a 70×70 square up into twenty-four squares, one of each size. To illustrate the sort of things that go wrong, try to explain why it is impossible to arrange thirty-one 2×1 dominoes on a chess board which has had two diagonally opposite corner squares removed. Such a dissection of the square would make a stunning, if rather difficult, jigsaw puzzle!

However, there are many dissections of a large square into squares that do not use a regular sequence of smaller squares. These kinds of dissections make excellent jigsaws, particular those which are squares made up of other squares, no two of which are the same. The smallest such dissection was first discovered by A. J. W. Duijvestijn and published in *Scientific American* in 1978. This consists of twenty-one squares with sides of length 2, 4, 6, 7, 8, 9, 11, 15, 16, 17, 18, 19, 24, 25, 27, 29, 33, 35, 37, 42 and 50. These fit into a square with sides of length 112. It is known that this is the unique dissection with twenty-one different squares and that there is no such dissection of a square into fewer squares. A jigsaw using Duijvestijn's dissection is shown in plate 15. A very similar puzzle could be made from a dissection by T. H. Willcocks that contains twenty-four different squares, using squares with sides of length 1, 2, 3, 4, 5, 8, 9, 14, 16, 18, 20, 29, 30, 31, 33, 35, 38, 39, 43, 51, 55, 56, 64 and 81. These all pack into a 175×175 square. The solution to this is shown in figure 6.2.

The smallest square in the dissection has sides with length one unit, and the smallest size convenient for a piece is 5 mm×5 mm, although even this can be difficult to handle. With this scale the largest piece is 405 mm × 405 mm, with the complete puzzle measuring 825 mm×825 mm. To appreciate how accurately the puzzle needs to be made, notice that if we draw a horizontal line through the centre of the square it crosses six pieces, whereas a vertical line through the fifty-five unit square cuts only three pieces. Evidently each piece must be made very accurately if the whole puzzle is to fit together in a satisfying way, without unseemly gaps—and do not forget that they must fit equally well in any of the four rotational positions and on both sides. Straight edges and right angles can be checked very accurately with an engineer's or carpenter's square, and their size with Vernier callipers.

The most attractive feature of Duijvestijn's dissection as a plan for a real puzzle is not so much that it requires the minimum number of pieces but that the ratio of the side lengths of the largest to the smallest is 25:1—smaller than other published dissections. This is of great practical significance, and taking the lengths of the sides of the smaller square to be 5 mm, the largest is 125 mm and the whole square is only 280 mm, an altogether more convenient size. Three millimetre plywood is probably the easiest material to use. A completely different technique from that used in section 1.2 was needed for making this dissection. Start by cutting squares slightly oversize and bring one side straight by carefully sanding, then, with the aid of an engineer's square, do the same to the second side. We need to do some accurate measuring for the next two sides and rather than use a ruler use Vernier callipers open to the exact size required. Still checking for squareness and straightness, test with the callipers: the size will be right when the jaws of the callipers just, and only just, drop across the square. Check the fourth side for squareness from both ends, and sand down to the point where the callipers drop: you will have one accurate square. It is best to start with the largest square so that if you end up under size it can be used for another piece. The final sizing is best carried out with wet and dry paper glued to a flat strip of wood. We have found that making three or four squares in one session is enough, as it does require extreme concentration. The result, however, is well worth the effort.

Notice that to complete both these patterns one takes the *greedy approach*. In terms of Willcocks's dissection, this means

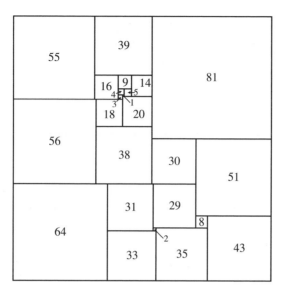

Figure 6.2. Willcocks's dissection.

that we begin with the largest two squares in opposite corners. Then we pack the next four largest to completely fill opposite edges. Next, the position of the 39-unit square should be clear, leaving a relatively small number left to fit in. However, the greedy approach is not always best. For example, assume we have five suitcases, four are 1×1 and one is slightly larger at 1.01×1.01. If we want to pack as many suitcases as possible into a 2×2 car boot, the greedy approach is clearly the worst, since only one will fit! When and when not to take the greedy approach is often far from obvious. Scheduling, ordering and packing problems are often mathematically very similar, and they are of huge practical importance.

6.3 Packing

In general, packing problems in mathematics are very difficult. By this, we do not mean difficult conceptually, but difficult in a computational way. Imagine you are going to use an exhaustive search method and look at every possible combination. There are just so many ways one could try to pack that the computation may effectively never finish. For example, imagine you want to arrange n different items in a queue. There are n ways of choosing the first. For each of these there are $(n - 1)$ ways for the second. Repeating this we see that in total the number of arrangements is

$$n \times (n - 1) \times (n - 2) \times \cdots \times 3 \times 2 \times 1 = n!.$$

The function $n!$ (pronounced 'n-factorial') grows very rapidly indeed: for example, $65! \approx 8.2 \times 10^{90}$, which is quite significantly more ways than there are particles in the known universe. Much more modestly, if you wanted to display all possible arrangements of only a dozen items, at the rate of one arrangement per second, it would take you $12! = 479\,001\,600$ seconds to do so, which is over 15 years.

One modern and exciting solution to this problem is to use strands of RNA (ribonucleic acid). These are complex molecules that react together to form long chains. The order in which this chain forms is random, and a test tube will contain some 10^{19} such molecules. This random linking is effectively like a parallel computation, with each combination being formed simultaneously. In many of these problems, recognizing a solution when you have it is comparatively simple. We can see when the parts 'fit' together. The problem is finding out whether or not a sequence corresponding to a mathematical solution is present at the end. Hence, if you look at which strands of RNA are produced in the reaction, you can find out whether a solution to your problem is there. This is done by a chemical reaction, very much like DNA fingerprinting. One of the pioneering papers in this new and exciting computational field is Adleman (1994).

6.4 Plane Dissections

In general, given any two polygonal figures of equal area, there exists a dissection between them using only a finite number of pieces. This result is proved in a number of stages, and we sketch an informal argument. The important observations are that (i) any polygonal figure may be divided into triangles and (ii) any triangle may be dissected and rearranged into a rectangle of *prescribed* base length. We can use a stack of such

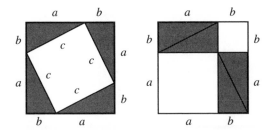

Figure 6.3. The Pythagorean theorem.

rectangles as intermediates between the two polygonal figures. Of course, such an argument has much detail to fill in, and in general this procedure will result in a large number of pieces. Perhaps a more interesting problem would be to find the *minimum* number of pieces needed for a dissection between two polygonal shapes. This problem appears to be very difficult and is currently unsolved in the general case. An extensive treatment of the history and mathematics of dissections is given in Frederickson (1997), with special hinged dissections covered in Frederickson (2002).

We cannot resist including another illustration of the Pythagorean theorem. In chapter 8 we will examine the problem of calculating area much more closely, and in the process give two more proofs. Collecting such proofs could become an obsession if one is not careful! For a collection of 365 more-or-less different proofs of the Pythagorean theorem see Loomis (1968). The illustration in figure 6.3 is satisfying because it is immediately obvious that the area of the unshaded region on the left is c^2, and the areas of those on the right are a^2 and b^2. The link between them is achieved by moving four copies of the original triangle itself. The figure on the left shows that $(a + b)^2 = c^2 + 2ab$. On the left we have $(a + b)^2 = 2ab + a^2 + b^2$. A wooden model of this is illustrated in figure 6.4.

It should be noted that modelling a geometric dissection always requires precise work, although the model shown in figure 6.4 was made without any measurement of length. The four triangles were cut from 1.5 mm plywood and finished clamped together to make sure they were all the same and were truly

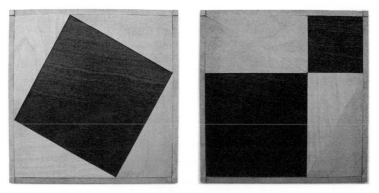

Figure 6.4. A model illustrating the Pythagorean theorem.

right-angled triangles with straight edges. The tray was made to fit, again without measurement.

6.5 Ripping Paper

We would like to propose another experiment in this section. You will need a couple of sheets of paper for this and you will be tearing these up into pieces.

For the first experiment take a square of paper and rip it in half. Take the remainder and tear this in half again. Continue this process forever, or at least until your remaining piece becomes too small, at which point you will be forced to let a thought experiment take over. Now, what you have strongly suggests that

$$\frac{1}{2} + \frac{1}{4} + \frac{1}{8} + \frac{1}{16} + \cdots + \frac{1}{2^k} + \cdots = 1. \tag{6.3}$$

This is an impression that figure 6.5 should reinforce.

Next take the other piece of paper and tear this into three equal bits and start to make two piles with two of the pieces. Take the third and rip this into three more equal size pieces and add one to each of the piles. Each pile now has

$$\tfrac{1}{3} + \tfrac{1}{9}.$$

Again, tear up the remaining bit into three and add one new bit to each pile. Keep doing this, and at each stage tear the paper

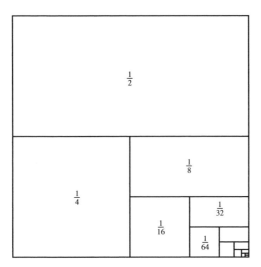

Figure 6.5. Infinite division of the square into successive halves.

you are holding into three pieces of equal size. Once the experiment is complete you will have two identical piles, each of which suggest that

$$\frac{1}{3} + \frac{1}{9} + \frac{1}{27} + \frac{1}{81} + \cdots + \frac{1}{3^k} + \cdots = \frac{1}{2}. \qquad (6.4)$$

You might like to sketch a counterpart of figure 6.5 that corresponds to this new experiment.

In each of these experiments we end up creating a pile of paper and then adding up the total area each piece represents. In the resulting sums, each term is a certain number of times the preceding term. In the first experiment we had a half and in the second a third. We could use any number as the multiplying factor and, for example, consider a sum such as

$$1 + 2 + 4 + 8 + 16 + \cdots.$$

All of these are known as *geometric sums*. Such sums have direct applications in many areas of mathematics, including the calculation of the total compound interest in a regular savings account.

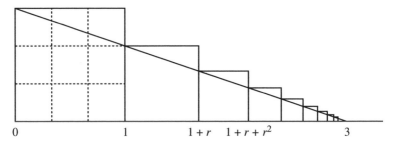

0   1 $1 + r$ $1 + r + r^2$ 3

Figure 6.6. Illustrating the convergence of a
general geometric progression, $r = \frac{2}{3}$.

In a general case the multiplier is r. For fixed $r \neq 1$ and n it is
a standard result that

$$1 + r + r^2 + r^3 + \cdots + r^{n-1} = \frac{1 - r^n}{1 - r}. \qquad (6.5)$$

When $r = 1$ the value of (6.5) is simply n.

This series is very important indeed and will give us our first
bite at infinity. When $-1 < r$ and $r < 1$ we can prove that r^n
tends to zero as n tends to infinity. That is to say, as n gets
larger and larger, the values of r^n eventually remain arbitrarily
close to zero. This allows us to make sense of the limit

$$1 + r + r^2 + r^3 + r^4 + \cdots = \lim_{n \to \infty} \frac{1 - r^n}{1 - r} = \frac{1}{1 - r} \quad \text{for } -1 < r < 1.$$
$$(6.6)$$

The above is an infinite sum, to which we have ascribed a definite
value, as was suggested by the experiment. Taking $r = \frac{1}{2}$ and
subtracting the first term gives us the famous result (6.3).

The illustration of the case $r = \frac{1}{2}$ in figure 6.5 is essentially
a special trick and does not generalize. However, the diagram
shown in figure 6.6, given by Casselman (2000), allows us to
see this for any $0 < r < 1$. The squares of side 1, r, r^2, etc.,
are simply placed side by side. Where the line that connects all
the top left corners intersects the baseline is the sum of the
progression. The case for $r = \frac{2}{3}$ is shown in figure 6.6.

Actually, forgetting the restriction $-1 < r < 1$ in the identity

$$1 + r + r^2 + \cdots = \frac{1}{1 - r}$$

is an excellent source of mathematical nonsense. For example, taking $r = 2$ gives

$$1 + 2 + 4 + 8 + 16 + \cdots = \frac{1}{1-2} = -1,$$

and taking $r = -3$ gives

$$1 - 3 + 9 - 27 + 81 - 242 + \cdots = \frac{1}{1+3} = \frac{1}{4}.$$

The modern student of analysis would immediately object that the partial sums do not converge; however, Euler (2000, section 109) does not dismiss their importance entirely.

> However, it is possible, with considerable justice, to object that these sums, even though they seem not to be true, never lead to error. Indeed, if we allow them, then we can discover many excellent results that we would not have if we rejected them out of hand. Furthermore, if these sums were really false, they would not consistently lead to true results; rather, since they differ from the true sum not just by a small difference, but by infinity, they should mislead us by an infinite amount. Since this does not happen, we are left with a most difficult knot to unravel.

6.6   A Homely Dissection

The final light-hearted dissection that we discuss can hardly be described as a mathematical model, or even a model, at all, but it does show that when no material is lost with a shearing cut by knife or scissors a more relaxed approach is possible. It can be demonstrated most easily on the breakfast table: it might even be construed as a lesson in table manners and be used in a book on etiquette. All that is required is a thin slice of bread cut from a square loaf and cut again diagonally to form a 90° isosceles triangle (see figure 6.7). Good manners dictate that it should be cut in half before attempting to eat it.

The most obvious cut is shown where D bisects AC. If AB and BC, the sides of the original loaf, are taken to have lengths of

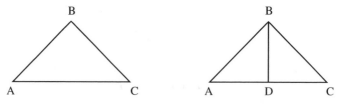

Figure 6.7. An obvious way to cut toast.

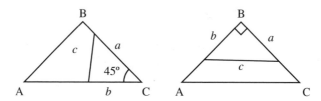

Figure 6.8. Other ways of cutting a slice of toast.

1 unit, then the total area is $\frac{1}{2}$, the length of the cut is $\frac{1}{\sqrt{2}} \approx 0.707$, and the area of each piece $\frac{1}{4}$.

The problem is to discover if there is a shorter cut that would still divide the slice into two equal areas, thus saving wear and tear on the plate and knife. However the cut is made, one of the 45° corners or the 90° corner will be in the triangular section and a quadrilateral will remain. Since the more promising solution looks to be to slice off the corner C (on the left-hand drawing) rather than B (on the right), we shall examine that possibility first (see figure 6.8). The area of the small triangle is

$$\text{area} = \frac{ab\sin(45°)}{2} = \frac{ab}{2\sqrt{2}} = \frac{1}{4}.$$

Hence $ab = \frac{1}{\sqrt{2}}$. Applying the cosine rule,

$$c^2 = a^2 + b^2 - 2ab\cos(45°), \tag{6.7}$$

gives

$$c^2 = a^2 + \frac{1}{2a^2} - 1.$$

Using calculus to find the maximum or minimum value we consider

$$2c\frac{dc}{da} = 2a - \frac{1}{a^3} = 0.$$

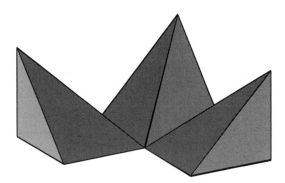

Figure 6.9. A dissection of a cube into three identical yángmǎ.

Therefore, $a^4 = \frac{1}{2}$ and so $a^2 = b^2 = \frac{1}{\sqrt{2}}$. Hence, the triangle in which c is shortest is isosceles. The length of cut is therefore approximately 0.6436, a saving of approximately 9% compared with the original 0.7071. While we have not proved that this is a minimum we leave it to you to consider the case when we cut off the 90° angle at B. A 9% saving is a poor return for the effort of cleaning butter from a ruler, and such an approach to cutting the bread is so unorthodox that it might itself be considered to be a breach of etiquette.

6.7 Something More Solid

Of course we can cut up solids into pieces, as well as flat objects into jigsaws. Dissections of solids are generally much harder in practice. One exception is the dissection of a cube into three identical square-base pyramids, known as yángmǎ, shown in figure 6.9. The three yángmǎ can also be hinged together. The net shown in figure 6.10 can be constructed very simply using the following origami construction, which is given in Brunton (1973). It is indeed fortunate that we have this origami-type construction, since $\sqrt{2}$ does not appear on many rulers, and measuring this length is necessary for the construction.

We start with a square ABCD, which surrounds the net. The procedure is as follows.

1. Fold along AC.
2. Fold DC on AC to find the point E on AD. (You should show that this point is in the correct place!)

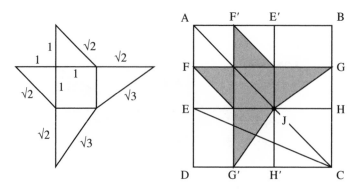

Figure 6.10. The net of a yángmǎ.

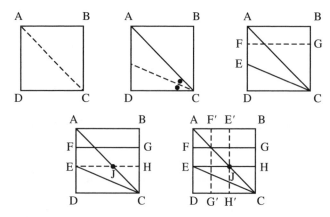

Figure 6.11. The construction steps for a yángmǎ.

3. Make a fold FG so that AB is perpendicular to AD through E.

4. Fold EH, parallel to DC. Where EH intersects AC mark on J.

5. Fold the two lines E'H' and F'G', both parallel to AD, as shown in the bottom right of figure 6.11.

Since three identical yángmǎ make a cube, their volume is one-third that of the cube. This result is actually considerably more general: the volume of any straight-sided pyramid is one-third of the base area multiplied by the vertical height. In chapter 8 we consider area in detail, and illustrate why the area of a circle is πr^2. Hence the volume of a circular cone must be $\pi r^2 \frac{1}{3} h$.

Chapter 7

FOLLOW MY LEADER

And so no force, however great, can stretch a cord, however fine, into a horizontal line that shall be absolutely straight.

Whewell (1819)

The principal characters in this chapter are two closely related transcendental curves, the tractrix and the catenary, with a strong supporting role being played by a conic section, the parabola. The ways in which the curves are formed are described by the Latin roots of their names: *tractare*, to pull or draw, and *catena*, a chain. The tractrix can be demonstrated most conveniently by pulling a weight across a horizontal table by means of a thread. The only way thread can exert any force on the weight is by traction, and in figure 7.1 if the weight starts from A_0 and the length of the thread is a, we can see that it can only be moved in a direction directly towards the 'tractor' travelling along what we might as well take to be the x-axis.

Its path is a tractrix, with the x-axis as asymptote. The tractor could just as well have moved in the opposite direction, hence the curve is symmetrical about the y-axis, and a tangent from the curve to the x-axis is of constant length. The following derivation of the Cartesian equation refers to figure 7.2. For future reference, note that $y/\sin(t) = a$, and so is constant.

The length of PB (the tangent) is a, and applying the Pythagorean theorem to the triangle PBC we have CB $= \sqrt{a^2 - y^2}$ and therefore

$$\frac{\mathrm{d}y}{\mathrm{d}x} = \frac{-y}{\sqrt{a^2 - y^2}},$$

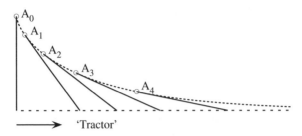

Figure 7.1. Pulling a weight with a thread.

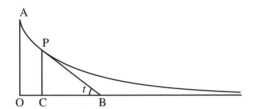

Figure 7.2. Half of the tractrix.

or

$$\mathrm{d}x = \frac{-\sqrt{a^2 - y^2}}{y}\,\mathrm{d}y.$$

This is a standard integral giving

$$\pm x = a \cosh^{-1}\left(\frac{a}{y}\right) - \sqrt{a^2 - y^2}.$$

The solid of rotation about the asymptote is known as a *pseudo-sphere*. A model of half of this solid is shown in plate 16. Intriguingly, such a shape has the same volume and surface area as a sphere of radius a. Its principal interest to engineers is that its shape is the ideal form for a thrust bearing, with the load along the axis. Ideally, the wear on the bearing should be even, in the sense that the whole curve is worn away and displaced to the left, by a small distance l, say. If we take a tractrix and displace it to the left by l, the component of the movement in a direction normal to the curve is then $l\sin(t)$, where t is the angle marked in figure 7.2. The work done by friction on some small area A is proportional to $Al\sin(t)$. In fact, the work against

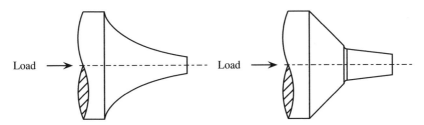

Figure 7.3. (a) The thrust bearing and (b) Schiele's bearing.

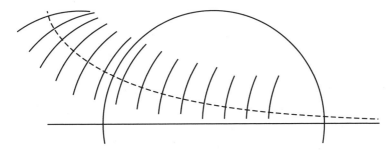

Figure 7.4. A tractrix as an orthogonal
trajectory through a set of circles.

friction is proportional to $Al2\pi y$ per revolution, where the coef-
ficient of friction and the axial loading are assumed to be constant. Therefore, to obtain our even-wearing bearing we require
y to be proportional to $\sin(t)$, and, as we have already seen,
this is true of the tractrix. In practice it is difficult to machine
both male and female components of a bearing to be tractrices.
Schiele approximated the curve by two straight lines producing a
two-cone bearing. Notice the discontinuity between the angles to
allow for easier grinding and polishing: the same technique was
mentioned earlier with the vee-blocks. Both the thrust bearing
and Schiele's bearing are shown in figure 7.3.

Returning to figure 7.2, it is clear that a circle of radius a
centred at B cuts the curve orthogonally—this provides an alternative way of describing a tractrix as the orthogonal trajectory
through a set of circles where the centres lie on a straight line.
The axis itself is also an orthogonal trajectory. This suggests
another way of drawing a tractrix. The more closely the circles
are spaced the more closely the short straight-line segments will

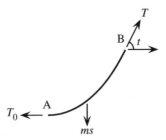

Figure 7.5. Forces on a section of hanging chain.

approximate to the true curve (see figure 7.4). Larger and more general tractrices are described by the path of the rear wheel when a bicycle is steered in different directions, and the same can be seen when a stretch limousine or long articulated lorry takes a sharp turn. The final example we give is of a lethargic dog being taken for a walk on the end of a long straight chain. This is just the same as pulling a weight if the dog walks slowly and the lead is always under slight tension. However, if you stop and the dog approaches you, the lead will slacken and form the next curve we describe: a catenary.

A catenary is the natural curve taken by a uniform flexible chain acted on only by gravity as it hangs between two supports. Its equation can be found by looking at half of the chain (see figure 7.5). Here, A is the midpoint of the curve, its lowest point, where the tension is T_0. At some other point B on the curve, the tension is T, making an angle t with the horizontal. We assume that the mass of the chain is m per unit length, and the arc length from A to B is s. We can now resolve forces:

$$T_0 = T\cos(t) \quad \text{and} \quad ms = T\sin(t),$$

giving

$$\tan(t) = \frac{ms}{T_0}.$$

Define $a = T_0/m$ so that $s = a\tan(t)$. This is the intrinsic equation of the catenary, being independent of any coordinate axes. For our purposes, though, it is more convenient to be able to

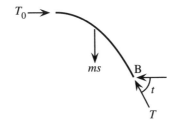

Figure 7.6. Forces on a section of an arch.

Figure 7.7. A suspension bridge.

write it in the more common and applicable form of the Cartesian equation in x and y:

$$\frac{dx}{dt} = \frac{dx}{ds}\frac{ds}{dt} = \cos(t)a\sec^2(t) = a\sec(t)$$

and

$$\frac{dy}{dt} = \frac{dy}{ds}\frac{ds}{dt} = \sin(t)a\sec^2(t) = a\tan(t)\sec(t).$$

These expressions may be integrated to give

$$x = a\ln(\tan(\tfrac{1}{4}\pi + \tfrac{1}{2}t)),$$
$$y = a\sec(t).$$

The elimination of t gives

$$y = a\cosh(x/a).$$

If we turn the diagram upside down to form an arch, the only changes are that what was in tension is now in compression and the directions of both arrows have been reversed. This leads to the conclusion that the inverted catenary is the natural, and best, form for an arch (see figure 7.6).

The main cables in a suspension bridge hang in a catenary, certainly before any decking is added, and this extra load has

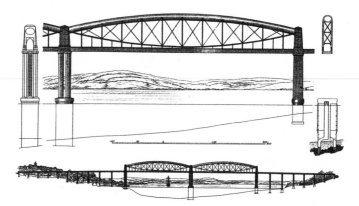

Figure 7.8. Brunel's Royal Albert Bridge
across the River Tamar at Saltash.

very little effect on its shape. In the crude sketch of figure 7.7
the two towers are being pulled towards each other because of
the unbalanced loading. The obvious solution, seen in virtually
all suspension bridges, is to extend the cables over the top of
the towers to massive anchorages at the ends of the approach
roads. This is possible only if the approach and centre spans
lie in a straight line. This solution was not available to Isambard Kingdom Brunel when designing the Royal Albert Bridge
to cross the River Tamar at Saltash. The lie of the land is such
that the approaches to the main spans over the river had to be
sharply curved, making it impossible to construct a conventional
suspension bridge. The elegant solution he proposed, and built,
was to use the catenary in two ways. One half of the load of the
rail deck was suspended by chains, as in a suspension bridge,
while the other half of the load was suspended from a beam in
the form of an inverted catenary. In this way the loading of the
three towers of the two main spans of the bridge was balanced so
that no staying cables were needed. This is shown in figure 7.8.

Returning now to the geometry of the curve, we consider the
evolute, which is the locus of the centres of curvature. The evolute of the tractrix is the catenary. This is shown in figure 7.9.
AB is a tangent to the tractrix at B and CB is a normal to it
whose length equals the radius of curvature at B. The locus of C
describes a catenary. In fact, there is no need to know the radius

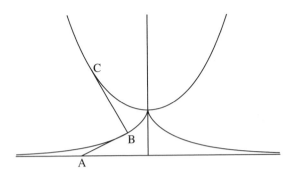

Figure 7.9. A catenary as the evolute of a tractrix.

of curvature as CB is a tangent at C so a series of normals will define the envelope of the same catenary. In the same way, the tractrix is the evolute of a catenary.

We can now introduce the parabola—an algebraic curve—and show how it is related to a catenary. The first point to note is that near their vertices the curves are indistinguishable. As we have already seen, a catenary is the curve taken by a uniform flexible chain, so its loading is constant per unit length of arc. A parabolic curve results when the loading is constant per unit *horizontal* length.

In order to show that this is true we can repeat the analysis already used to derive the equation of the catenary with only a few small but significant changes. The principal change is that the curve has been drawn on Cartesian axes, and the loading is now m per unit horizontal length; $\tan(t)$ can be written dy/dx. After resolving forces,

$$\tan(t) = \frac{mx}{T_0} = \frac{dy}{dx}.$$

Assuming that the curve passes through the origin, a solution is $y = x^2/4a$, for a suitable choice of constant a. This curve is a parabola.

In a suspension bridge where neither the weight of the cable nor that of the decking can be neglected, the actual curve must be somewhere in between. The fact that the curves are very similar, almost indistinguishable, near their vertices is due to the fact that they are tangential to the horizontal. Here, the ratio between

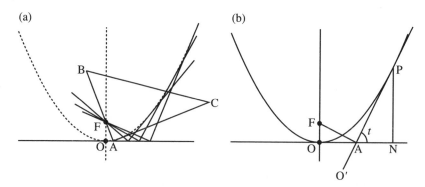

Figure 7.10. Generating a parabola from an envelope of lines.

uniform horizontal loading and uniform loading per unit length is proportional to the cosine of a small angle, which is extremely close to one.

A more fundamental relation between the two curves is that the locus of the focus of a parabola rolled along a straight edge is a catenary. This may appear strange at first sight because one is an algebraic curve while the other is transcendental. The link arises because the expression for the arc length of a parabola is given by a transcendental function: the logarithm. In order to simplify what follows we look first at a parabola. One way of drawing this curve, or rather its envelope, is shown in figure 7.10(a).

Take a straight edge and fix this to a drawing board and place a pin, F, a distance a above the edge to act as the focus of the parabola. As the 90° corner of a square is moved along the edge with side AB resting against the pin, AC is tangent to the parabola. For the purposes of cutting out a template an envelope is just as easy to work to as a single line obtained from plotting the curve.

While you have a straight edge and pin to hand a brief digression is permissible. What would happen if you used, say, the 30° or 60° corner of the square instead of the 90° corner? If the straight edge is replaced by a line, and the square by a template of an angle greater that 90°, what is the outcome? Defining curves as the envelope of a family of lines provides some very interesting and enjoyable problems. You can also make some curves yourself with card and thread, and these can make very interesting decorative patterns. We shall again make extensive use of curves defined as the envelope of a family of lines in section 10.9.

In figure 7.10(b) the parabola $x^2 = 4ay$ is shown with the tangent from P, which cuts the x-axis at A, and where PN is drawn perpendicular to the x-axis. With the focus at F, join F and A. By definition, angle FAP equals $90°$. In the triangle PNA we have

$$\frac{dy}{dx} = \frac{PN}{AN},$$

and hence AN $= \frac{1}{2}x$. Therefore, OA $= \frac{1}{2}x$ and so

$$PA = \sqrt{PN^2 + AN^2} = \sqrt{y^2 + \tfrac{1}{4}x^2} = \tfrac{1}{2}x\sqrt{1 + \frac{x^2}{4a^2}}.$$

In the triangle FOA we have

$$FA = \sqrt{a^2 + \tfrac{1}{4}x^2} = a\sqrt{1 + \frac{x^2}{4a^2}}.$$

Another way of looking at the diagram is that the parabola is stationary and the straight edge has rolled round until P is the point of contact. The origin O has now moved to a new position O'. The length of FA is then the new y-coordinate of the focus, say Y. The length of O'P must be equal to arc length OP. Calculating this arc length is a rather lengthy calculation, which we have chosen to omit since it adds little to our story. It is quite standard and can be found in many calculus books such as Maxwell (1954, volume II, p. 105). The result we need is that

$$O'P = \tfrac{1}{2}x\sqrt{1 + \frac{x^2}{4a^2}} + a\ln\left(\frac{x}{2a} + \sqrt{1 + \frac{x^2}{4a^2}}\right).$$

Given that X $=$ O'P $-$ PA we have that

$$X = a\ln\left(\frac{x}{2a} + \sqrt{1 + \frac{x^2}{4a^2}}\right) = a\sinh^{-1}\left(\frac{x}{2a}\right).$$

Figure 7.11. Jefferson National Expansion Memorial. Photo courtesy of the National Park Service, Jefferson National Expansion Memorial.

Or $\sinh(X/a) = x/2a$ and

$$X = a \ln \left(\frac{x}{2a} + \frac{Y}{a} \right),$$

therefore $e^{X/a} = (x/2a) + (Y/a)$. From this we have

$$\frac{Y}{a} = e^{X/a} - \sinh \left(\frac{X}{a} \right) = \tfrac{1}{2}(2e^{X/a} - e^{X/a} + e^{-X/a}) = \cosh \left(\frac{X}{a} \right),$$

which proves the relationship between the two curves.

The grand finale to this chapter is a catenary arch, the Gateway Arch, a striking part of the Jefferson National Expansion Memorial at St Louis, Missouri, in the United States. This stainless steel monument stands 630 feet high and is 630 feet wide at ground level. Its cross-section is that of an equilateral triangle whose side length tapers from 54 feet at the base to 17 feet at the top, where there is an observation room.

The form of the arch is defined by

$$y = 68.7672 \left(\cosh \left(\frac{x}{99.668} \right) - 1 \right) \text{ feet.}$$

This is not the equation of any curve seen by just looking at the arch, e.g. the outside boundary seen in figure 7.11. It is the more important equation defining the locus of the centre of gravity of the cross-sections of the tower. The derivation of the equation of a catenary given earlier in the chapter assumed a constant mass per length of curve, although this is certainly not the case with this arch.

Chapter 8

IN PURSUIT OF COAT-HANGERS

Mathematical Instruments are the means by which those Sciences are rendered useful in the Affairs of Life. By their assistance it is that subtle and abstract Speculation is reduced to Act. They connect as it were, the Theory to Practice, and turn what was bare Contemplation, to the most substantial Uses.

Stone (1753)

In previous chapters we have examined how to draw a straight line and the problems concerned with marking a ruler. In each case we discovered some subtle features associated with these simple ideas. Now we turn our attention to the problem of ascertaining the area of a plane shape.

Area can certainly be measured with complex scientific instruments, such as those shown in figure 8.19 or in figure 8.25. Seeing the undoubted precision involved in their manufacture it may come as a surprise that we are able to begin this description of a *hatchet planimeter* with details of how to make one. Furthermore, the required raw materials and tools are found in most homes and there is no need for the facilities of a fully equipped workshop.

Start with a metal coat-hanger, although any soft wire will do equally well. Cut off a length of about 200–300 mm. Flatten one end to form the hatchet, taking care all the time to keep it symmetrical. A hammer and something to serve as an anvil are all that you need. Clean it up with a smooth file to give a nicely rounded knife blade. *This should not be sharp enough to cut anything.* The other end is filed to a blunted point with a smooth tip that will not scratch paper. Now bend this into the form shown so that the whole lies flat in a plane and the knife

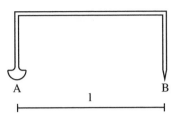

Figure 8.1. A simple hatchet planimeter.

edge is in line with the point. The easiest way to check this is to draw the point along a straight line and see if the knife edge deviates from it. The direction of the knife blade can be tweaked using pliers if necessary. The length of the horizontal distance, marked l in figure 8.1, does not matter, but you will need to know what it is. A ruler is sufficiently accurate to measure this to the nearest millimetre, and 100 mm is a good length to aim for.

The last job is to glue (or tape) a couple of washers or a nut and a bolt or even obsolete coins above the blade at A. This is not shown in figure 8.1 but is in figure 8.2. This adds a little mass above the knife blade and makes the device much easier to use in practice—the weight need not exceed 200 g.

You have made a *hatchet planimeter* and although you may think it looks crude and certainly nothing like a precision scientific instrument capable of accurate work, you are wrong. The prototype was made by a Danish blacksmith for its inventor Holgar Prytz.

To measure the area of an irregular plane shape estimate by eye where its centre of mass is and mark it. With a straight edge, draw a line through this point, extending it beyond the centre by at least a distance l. With the pointer B on the centre record the position of the knife edge A on the line by pressing it down lightly to make an indentation. Hold the planimeter loosely at B and drag the pointer along the line to the periphery, once round the region and finally back to the centre again. Mark the new position of the knife edge with another indentation. The displacement of the edge multiplied by the length of the planimeter l gives a very good approximation to the area.

When using the planimeter it is important to keep the whole device vertical and guide the pointer at B without applying any

Figure 8.2. Using a hatchet planimeter.

twisting force. It is vital that A does not slip sideways, only moving forwards and backwards in pursuit of B. The reasons for this will become apparent when we explain how it works.

What is the appropriate relation between the length of the planimeter and the maximum width of the shape whose area is to be measured? This is a practical question, and the general rule is that the maximum width of the shape should not exceed $\frac{1}{2}l$, but this is a matter of compromise. The intrinsic mathematical error is reduced the greater the ratio l/width is, but the penalty is that the proportional error in measuring the displacement is increased. If the width is too large, one option is to divide the shape into two or more parts and measure each separately. Conversely, if the shape is too small, trace around it twice before measuring the displacement.

The errors inherent in this planimeter will be discussed later. For now, it is interesting to start with a simple shape whose area is known. Try measuring the area of a square of diagonal width $\frac{1}{2}l$ or a circle of diameter $\frac{1}{2}l$. The positions of the centres of mass are known, so start from there and then find out what happens if this starting point is deliberately misjudged. Record your results to get some feel for the accuracy of your planimeter and your own skill in using it.

It should be noted that the need to find the area of a plane shape is very common indeed, particularly in surveying and cartography. There are many other engineering applications where there is a need to calculate the area under a graph. For example, in calculating the work done by a steam engine a reading of pressure within the cylinder over time is drawn using what is known as an *indicator*, which we examined in section 3.2. The graph is drawn from a direct machine reading and the area under this graph needs to be measured accurately. For this to be practical it should be simple and quick.

The key to the operation of the hatchet planimeter is the observation that as the pointer traces around the curve, the knife blade follows in what is termed a pursuit curve. Recall that the tractrix of chapter 7 is an example of a pursuit curve. The blade does not slip sideways, but only moves forwards and backwards. This is why care is needed not to twist B and where some mass above A really helps. Actually, the theory behind this type of planimeter shows us that the measurements obtained give only an approximation to the area. Of course, there will be practical errors in any measurement, but those are different. There are planimeters that mathematically guarantee to measure the exact area and we shall describe two of these as well, together with the source of mathematical error in the hatchet planimeter.

8.1   What Is Area?

In order to understand how a planimeter works we really need to discuss what area actually is. In a similar way that length quantifies the extent of a one-dimensional line, area quantifies the space occupied by a two-dimensional plane shape. We usually only consider the lengths of straight lines, but many curves can in fact be measured by using a wheel to roll along the curve. The length of a straight line with the same total roll as the curve is said to be the length of the curve. This idea is used in a map measurer, which can be used with an appropriate scale to find the length of a proposed journey. Area is defined in terms of length using the following simple rules.

1. The area of a rectangle is the product of the lengths of the sides. So, the area of the rectangle in figure 8.3 is $a \times b$.

2. If we cut a shape into parts, the area of the original shape is the sum of the areas of the parts.

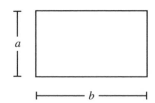

Figure 8.3. A plane rectangle.

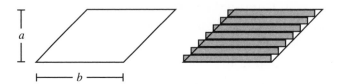

Figure 8.4. Dividing a parallelogram into small rectangles.

The first of these rules is the very basis of area and the second seems perfectly reasonable. However, with the second are you sure it does not matter *how* you decide to cut a shape up? It is a consequence of the second that the example in section 9.6 is impossible. Many of the properties that we expect from area are further consequences of these simple rules. For example, area is unaffected by rotation or translation in the plane. This means that we can move shapes around without changing their areas. The second rule holds even if one cuts a shape up into an infinite number of parts. For example, look back at figure 6.5. This shape is dissected into an infinite number of pieces and yet these sum to a finite area. The possibility for an infinite division gives area some very interesting and counter-intuitive properties.

Using these rules we will illustrate how to calculate the areas of various common shapes. Since area is based on rectangles, we shall start by using rectangles to calculate the areas of some simple and well-known shapes. The first of these is the parallelogram, which can be modelled very effectively with blocks or identically sized books.

Let us take a parallelogram, as shown in figure 8.4. We shall approximate this area by n thin rectangles of width b and height a/n. Immediately we notice that this approximation is not perfect. Small triangles stick out on one side and other small

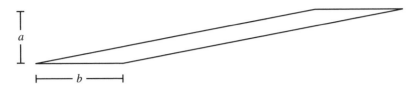

Figure 8.5. A stretched parallelogram.

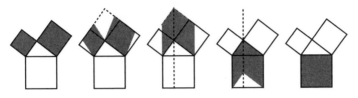

Figure 8.6. The Pythagorean theorem via parallelograms.

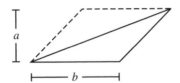

Figure 8.7. Area of a triangle via parallelograms.

triangles are missing on the other. In this case these errors cancel out, although we certainly cannot take this for granted. What is important is that as n increases, which is to say as we take more and more strips, we get a better and better approximation to the parallelogram.

This model is easy to assemble from n identical blocks. Once assembled, slide them horizontally until they are all one above each other. Since area is preserved by translation in the plane we immediately see that the area of the parallelogram is equal to the rectangle of width b and height a. This means that however far we slide the blocks, the area of a parallelogram remains constant. The parallelogram in figure 8.5 has the same area as those in figure 8.4. Manipulations of parallelograms can be used to provide another demonstration of the Pythagorean theorem, which is shown using the five pictures in figure 8.6. Notice the use of shears and translations to turn squares into parallelograms, move them, and then shear them back into a single square.

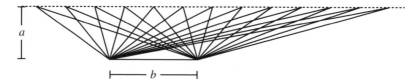

Figure 8.8. Moving the vertex of a triangle parallel to the base.

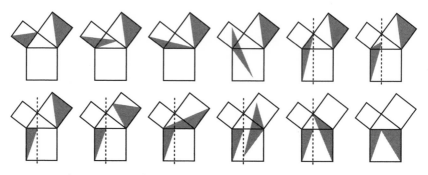

Figure 8.9. The Pythagorean theorem via triangles.

The next simplest shape is the triangle, such as that shown in figure 8.7. This can be seen as sitting inside a parallelogram which in fact contains two identical triangles. Therefore, the area of a triangle is half the product of the base b and the height a. That is $\frac{1}{2}ab$. Again, we may move the apex of the triangle in a way that is parallel to the base without altering the area. All the triangles in figure 8.8 have identical area.

The previous demonstration of the Pythagorean theorem relied on shears and translations of parallelograms. The actual proof given by Euclid in book I, proposition 47 is different, and relies on shears and rotations of triangles. What he does is prove that *half* of the areas of each of the squares on the two sides combine to give half the area of the square on the hypotenuse. The argument given by Euclid relies on labelling the various points and justifying each step carefully using previously proved propositions. This proof amounts to the sequence of pictures in figure 8.9.

Another vitally important mathematical shape is the circle. Next we shall use shears, translations and rotations of triangles to show that the area of the circle of radius r is πr^2. To do

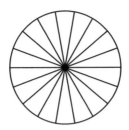

Figure 8.10. A circle divided into sectors.

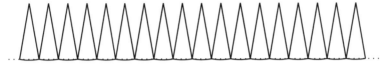

Figure 8.11. Sectors of a circle arranged on a line.

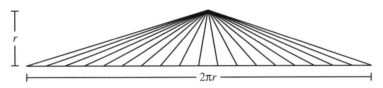

Figure 8.12. Gather the sectors to form a triangle.

this, begin with a circle of radius r, which by the definition of π has circumference $2\pi r$. This circle is divided up into n identical sectors (see figure 8.10). Next we cut up the circle into sectors and arrange these along a straight line as shown in figure 8.11. This does not affect the area. Now, the length of this line is, for large n at least, approximated very well by the circumference of the circle. Furthermore, the height of each sector is practically the radius of the circle. We shall approximate the area of each sector by such a triangle. You would be well advised to retain a suspicious attitude towards the cavalier way we forge ahead with approximations such as these. This notwithstanding, we can now move the apex of each triangle horizontally and gather these together in one place (see figure 8.12). Such a displacement of the apex does not alter the area. The area of this large triangle is clearly πr^2. Remember that this triangle is formed from the circle cut up into sectors, rearranged and with the individual triangles gathered together. Thus, we have the area of the circle also being πr^2.

It was Archimedes of Syracuse (287–212 B.C.E.) who first found this formula for the area of the circle and he gave it in his work *Measurement of the Circle*. He used a much more careful argument involving two sequences of regular polygons, one inside and one outside the circle. These sequences had an increasing number of sides and so approximate the circle with successively greater accuracy. In this way the circle is trapped between the inscribed and circumscribed polygons. It is possible to calculate the area of each polygon, so we also have two sequences of areas. These sequences both converge to the same quantity which must also be the area of the circle.

Many shapes can be decomposed into rectangles, triangles, circles and other regular shapes, and the areas of these shapes can therefore be calculated by summing the areas of their component parts. But what of an irregular shape, such as that shown in figure 8.2? How can we measure, or better still calculate, the area of such a shape? This problem appears to be much more difficult.

To illustrate why area is a difficult subject we give an example that begins to show how subtle the concept of area actually is. This shape, known as Koch's snowflake curve, is constructed in stages, starting with an equilateral triangle with sides of length 1. Next construct a *star of David* shape, which is made by putting an equilateral triangle of side length $\frac{1}{3}$ onto each of the sides of the original triangle. This is shown in figure 8.13.

We extend this idea to construct a sequence of progressively more complicated shapes by putting an equilateral triangle onto the middle third of each of the sides. If this is continued indefinitely, the limiting shape is known as *Koch's snowflake*. The first few stages in the construction are illustrated in figure 8.13.

The first shape has perimeter $P_1 = 4$. At every stage each side is replaced with four short sections, each one-third the length of the side they are replacing. Thus, if P_n is the perimeter of the nth stage, then

$$P_{n+1} = \tfrac{4}{3}P_n.$$

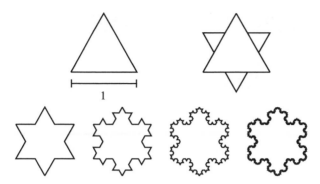

Figure 8.13. Koch's snowflake curve.

An induction argument gives

$$P_n = 3(\tfrac{4}{3})^n.$$

This sequence increases indefinitely and becomes unbounded. So the limiting shape has an infinite perimeter.

As for the area, it is clear that this will fit into a large circle and hence is finite. However this is not good enough and we prefer to calculate it exactly. We noted in (1.1) that an equilateral triangle of side l has area $\frac{\sqrt{3}}{4}l^2$. We started with $l = 1$, and at each stage we add $3 \cdot 4^{n-1}$ small equilateral triangles, each with sides of length $1/3^n$. The total area added gives the formula for the area A_{n+1} as

$$A_{n+1} = A_n + \frac{\sqrt{3}}{4}\frac{1}{3}\left(\frac{4}{9}\right)^{n-1}.$$

Solving this, we have

$$A_n = \frac{\sqrt{3}}{4}\left[1 + \underbrace{\frac{1}{3}\left(1 + \left(\frac{4}{9}\right) + \left(\frac{4}{9}\right)^2 + \cdots + \left(\frac{4}{9}\right)^{n-2}\right)}_{\text{this is a geometric progression, as in (6.6)}}\right].$$

Using (6.6) with $r = \frac{4}{9}$ we see that the total area of the limiting shape is

$$A_\infty = \frac{2\sqrt{3}}{5}.$$

Thus, by constructing Koch's snowflake we have produced a shape which encloses a finite area but which has an infinite

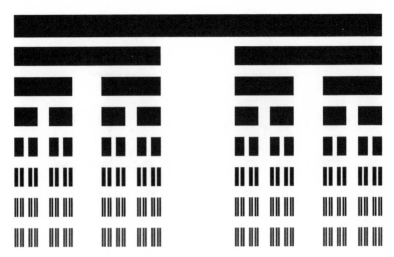

Figure 8.14. Cantor's set.

perimeter! In fact such shapes appear to be excellent mathematical models for naturally occurring shapes such as coastlines, fractures and clouds.

Area has some even stranger properties that the following example will help to reveal. We start with a solid filled-in square with sides of length 1. We shall remove vertical strips from this shape using the following scheme. We begin by removing a vertical strip of width $\frac{1}{5}$. From the remaining two strips we remove two strips of width $\frac{1}{25}$. This leaves four strips and from each of these we remove strips of width $\frac{1}{5^3}$. This process is made to continue forever and at each stage we remove 2^{n-1} strips of width $\frac{1}{5^n}$. Hence, the total area removed at each stage is

$$2^{n-1}\frac{1}{5^n} = \frac{1}{5}\left(\frac{2}{5}\right)^{n-1}.$$

If we continue this process indefinitely, it is clear that we will end up with a loose bundle of fibres. It is not clear that anything will remain; however, if two points with different x-coordinates are left, they will certainly end up on different fibres. The first seven stages of such a construction are shown in figure 8.14, in which the vertical height has been squashed severely for clarity.

In fact, the total area of material removed after n stages is only

$$\frac{1}{5}\left(1 + \frac{2}{5} + \frac{4}{25} + \cdots + \left(\frac{2}{5}\right)^{n-1}\right),$$

which is another geometric progression. Using (6.6) again, this time with $r = \frac{2}{5}$, we see that the total area removed is only

$$\frac{1}{5}\frac{1}{1 - \frac{2}{5}} = \frac{1}{3}.$$

So the remaining bundle of fibres has an area of $\frac{2}{3}$, even though it consists of an infinite collection of disjoint widthless vertical strips. Actually, we can repeat this process again and remove horizontal strips in exactly the same pattern. That is to say, beginning with a solid unit square, at each stage we shall remove 2^{n-1} horizontal and vertical strips each of width $\frac{1}{5^n}$. In doing this we shall remove at most a third of the remaining area leaving at least a third of the original area behind. However, this is now like a dust since any two distinct points remaining will be separated by previously removed vertical or horizontal strips or both. So a very fine and totally disconnected dust-like set of points can have a positive area!

8.2   Practical Measurement of Areas

Calculus is often encountered as a set of manipulations for doing such things as 'differentiation' and 'integration'. Using integration to find the area under a graph is elegant, powerful and accurate. These techniques can be extended to other shapes in the plane where the boundary is described by a tidy mathematical equation. What about the data obtained when measuring the power output during the stroke of an engine cycle? This kind of experimental data certainly cannot be described *exactly* using perfect mathematical curves. Yet can we still measure the area under a graph such as this with mathematical accuracy?

The experiment at the beginning of the chapter claimed to relate an arbitrary plane area to a measurement obtained by tracing around the perimeter. Certainly this would be enough to measure the area under a graph: just trace around it. But what connection, if any, exists between area and perimeter? That a circle encloses the largest possible area for a fixed perimeter is a famous result known as the isoperimetric theorem. In the other direction, squashing a circle shows that a given perimeter may enclose a very small area. Next take a fixed area. Adapting the construction used for figure 8.13 demonstrates that there is no upper limit on the perimeter which can enclose this. Furthermore, a collection of separate widthless fibres, or a collection of specks of dust, can possess a significant positive area! Hence, there appears to be little or no connection between area and perimeter.

Given these subtleties, planimeters are extremely surprising. Note that we did *not* measure the perimeter in the experiment, we only traced around the boundary. This physical movement induced a displacement in the hatchet blade. So that these devices work we shall restrict ourselves to plane shapes: those that are a single connected piece, with no holes, and which have a smooth boundary, except at a finite number of corners. This removes the pathological examples such as those of figure 8.13 and similar monsters but, since a smooth boundary can wiggle an awful lot, we are still left with a particularly surprising, not to mention useful, result.

In fact, we shall discuss three different planimeters. The first is known as the linear planimeter, an example of which is shown in figure 8.19. The second kind is known as the polar planimeter or sometimes as the Amsler planimeter after its inventor. This was the most successful and common kind and many examples survive. Indeed, this type of planimeter may well still be in use. An example of such a planimeter is shown in figure 8.25, and an enlargement of part of it is shown in figure 8.26. The third planimeter is known as the hatchet planimeter or sometimes the Prytz planimeter, again after its inventor. After the complexity of the polar and linear planimeters it is disarmingly simple. However, the mathematics of the hatchet planimeter assures us only of an approximation to the area, whereas the mathematics of the polar and linear planimeters guarantees an exact result.

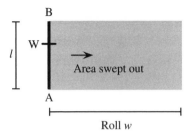

Figure 8.15. The area swept out by a line.

The basic operation of all planimeters is the same. To obtain a direct measurement of the area of a plane shape trace around the perimeter once in a clockwise direction. On the linear and polar planimeters the area is then read directly from a graduated counter that is part of the device. With the hatchet planimeter a separate ruler is needed. A comprehensive study of the types of planimeters available prior to 1894 can be found in Henrici (1894). The following geometrical explanations are based in part on this report.

8.3   Areas Swept Out by a Line

To start our discussion of the planimeter, consider a line AB of length l which moves a distance w to the right. As it moves, this line will sweep over a rectangle of area lw (see figure 8.15).

Suppose that a graduated wheel W is fixed using this line as an axle. As the line moves, the wheel will roll along. The total 'roll' of this wheel, given appropriate graduations and no slippage, will be the length w, just as with a map measurer. If the wheel is calibrated with l in mind, the area swept out may be read directly off this wheel.

In effect we have constructed a very simple planimeter, although since we can only measure areas of rectangles of length l this is practically useless. If the rod now moves vertically, no area will be swept out and the wheel will not roll. Again, the roll of the wheel gives the area swept out. Now consider the line moving diagonally across a parallelogram. The crucial observation is that the wheel will only record the component of the motion in the direction perpendicular to the line AB. As we showed above, the

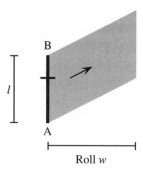

Figure 8.16. Moving the line across a parallelogram.

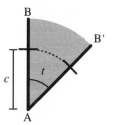

Figure 8.17. The rotating line.

area of such a parallelogram is the product lw. Again, the roll as recorded on the wheel multiplied by the length gives us the area swept out by the line (see figure 8.16).

Note that we have taken movement to the right to be positive roll. Similarly we take movement to the left to be negative roll and so consider the areas generated by a moving-line segment to be *oriented*. This means that we give area either a positive or a negative sign, and when adding such areas they can cancel each other out. Next consider the situation when we fix one end of the line at A and rotate the rod an angle t from B to B' say (see figure 8.17). The roll recorded on the wheel will be the arc length $w = ct$. The area swept out by the rotating line will be

$$\text{area} = \tfrac{1}{2}l^2 t = \frac{l^2}{2c}w$$

(as usual in mathematics we measure t in radians). Note that the area swept out is again proportional to the roll and also to the distance of the wheel from one end of the line.

In general, a line segment moving in the plane will sweep out an area and the direction of motion will dictate whether this area is considered positive or negative. Such a motion will contain both rotation and translation components. The fact that the three planimeters we discuss below work is justified by considering the areas swept out by moving-line segments. Of course, the justification above only applies to shapes with length l, so this is not much use. We shall remove this restriction in the next section.

Before we give an explanation of the mathematics behind the planimeter we comment on the wheel actually used on linear and polar planimeters. It is mounted on a short axle which is parallel to but slightly offset from AB. The wheel is connected by a worm gear to a counter and the wheel itself is read using a Vernier scale (discussed in section 4.5). The mechanism from an Allbrit 6000 planimeter is shown in figure 8.26. On this particular model the whole wheel mechanism can be moved along the line AB and positioned with the aid of another Vernier scale. This is to change a constant of proportionality, effectively changing l, so that when correctly set a direct reading of area can be taken in square centimetres, square inches or from a variety of scale diagrams. A detachable magnifying glass, not shown, is provided to help in accurate reading of the scales.

8.4   The Linear Planimeter

The linear planimeter incorporates the line AB, just as above, with a wheel at W. This time the point at A is constrained to run in a straight line, perhaps in a track or groove, such as that in the schematic of figure 8.18. Although this design was not as popular as the polar planimeter that we shall discuss in detail in section 8.5, commercial devices were based on this configuration (see, for example, Snow 1903). More popular commercially were designs that mounted a pivot for point A on a trolley, again constraining A to move in a straight line. This overcomes the difficulties with large maps and plans where a track, such as that shown in figure 8.18, gets in the way of B and so obscures part of the area to be measured. An example of such a trolley is shown in

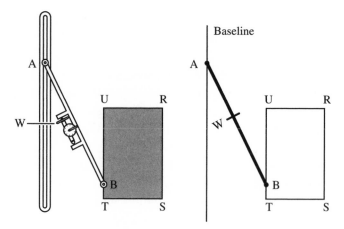

Figure 8.18. The linear planimeter.

Figure 8.19. An example of a Koizumi linear roller planimeter.

figure 8.19. Here the trolley moves in a straight line and the point A can be found at the end of the arm that is perpendicular to the axle. The point B can be seen as a small circle in the magnifying glass and the wheel is enclosed in the rectangular box, together with a counter.

To begin in a very simple way we shall measure the area of the arbitrary rectangle RSTU, shown in figure 8.18. We shall assume that the point B traces around the perimeter of this rectangle in a clockwise manner and that the point A is constrained to move in a line parallel to UT. Note that the dimensions of the rectangle are now arbitrary and the only restriction is that the length l, of the arm AB, must be sufficiently long to trace around the rectangle with A running in the track.

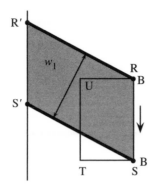

Figure 8.20. The motion of B from R to S.

During this motion, which is assumed to begin at R, B moves from R to S, then from S to T, and so on until B returns to R. We consider in detail the relationships between the roll as recorded on the wheel W and the area swept out by the line AB, and show that the total roll is proportional to the area exactly as before.

Consider first the motion of B along the edge from R to S. Since this is parallel to the track on which A is constrained, A will move in the same direction and it will move the same distance. During this motion the line AB sweeps out the shaded area shown in figure 8.20, and the wheel records a roll that is proportional to the normal component of the motion. That is to say, perpendicular to the line AB. This is the distance w_1 shown in figure 8.20. Since w_1 is the height of the parallelogram RSR'S' and the base is the length AB, which is l, the area swept out by this motion is lw_1.

Next the point B moves from S to T and, during this motion, the point A moves vertically, as shown in figure 8.21(a). For now we denote the roll as w_2. Next, assume that B moves from T to U. Once again, the line AB is translated and sweeps out a parallelogram, just as during the motion of B from R to S. This time the roll w_3 will be in the opposite direction and will therefore be negative. The area swept out will be $-lw_3$. This is shown in figure 8.21(b).

Lastly, we move B from U back to the initial position R. This roll is denoted by w_4. Notice that the movement from U to R is

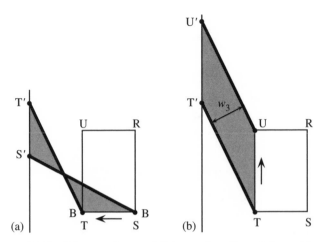

Figure 8.21. The motions of B from S to T and from T to U.

exactly the opposite of the movement from S to T but translated vertically. Thus, by appealing to symmetry, we see that the roll w_2 must be equal to $-w_4$. Thus, the total recorded roll must be

$$w_1 + w_2 + w_3 + w_4 = w_1 + w_3.$$

If we multiply this reading by l, the length of AB, we have

$$lw_1 + lw_3 = lw_1 - (-lw_3) = \text{area}(\text{RSR}'\text{S}') - \text{area}(\text{TUT}'\text{U}').$$

Exactly as before we may slide the base of the parallelogram RSR′S′ along the baseline, along which A is constrained to move, without changing its area. If we transform both parallelograms RSR′S′ and TUT′U′ into rectangles, we can more easily see that the difference between these two areas is exactly the area of RSTU shown in figure 8.22. This shows that, by moving B round RSTU and multiplying the total roll recorded on W by the length of AB, we have the area of RSTU.

Of course, we have only succeeded in showing that our planimeter works for certain rectangles. Next we consider a compound shape consisting of two joined rectangles, as shown in figure 8.23. If we trace around the perimeters of RSTU as before and then round XTVW, we will be able to calculate the total area. Notice that in doing this we trace along the line TX twice, but in

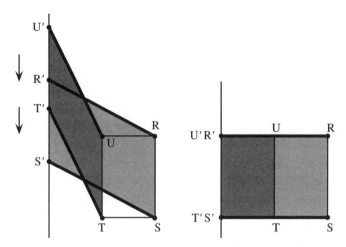

Figure 8.22. Translating the bases of the two parallelograms
leaves their oriented areas unchanged.

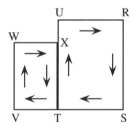

Figure 8.23. Two adjacent rectangles.

opposite directions. So the rolls recorded along this line cancel
each other out. This leaves a total roll exactly the same as if we
traced round the perimeter of the compound shape. Hence the
total roll recorded by tracing the perimeter is exactly the roll
recorded by tracing the two shapes separately.

Assume that a complicated shape can be cut up into adjoin-
ing rectangles. If we trace around each one separately, the roll
along interior lines will cancel leaving us with the roll round the
perimeter. Thus the area will be this reading multiplied by l. The
only difficulty is when a shape with a smooth curved boundary,
which cannot be dissected into a finite number of rectangles,
needs to be considered: see, for example, figure 8.24. The usual
approach is to say that 'infinitely small' rectangles make up the

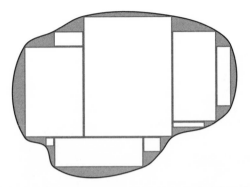

Figure 8.24. Tracing round a smooth boundary.

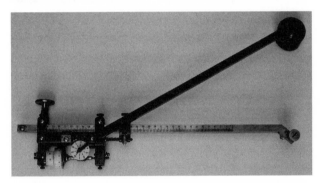

Figure 8.25. An example of a polar planimeter.

smooth shape, and so, strictly speaking, our argument breaks down here. However, in the case of a smooth boundary our planimeter still works. In fact we can cope with a finite number of kinks in the boundary but not a shape like the limiting form of the sequence generated by figure 8.13, which is decidedly unsmooth! In the case of a smooth, or finitely kinky, boundary the area will be found by reading the total roll and multiplying by the length l.

8.5   The Polar Planimeter of Amsler

The polar planimeter is very similar to the linear planimeter considered in the previous section. This design was first invented by Jacob Amsler (1823–1912) in the 1850s. This type of planimeter consists of two hinged arms, with the line AB as before, but this

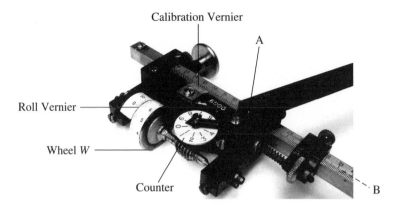

Figure 8.26. Details of the scales from an Allbrit 6000 planimeter.

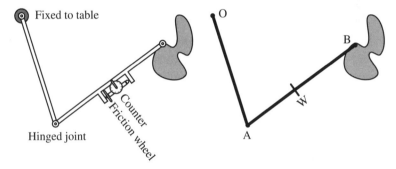

Figure 8.27. The polar planimeter.

time A is fixed to another arm and so is constrained to move in a *circle* rather than a straight line. A schematic is shown in figure 8.27 and an actual planimeter in figure 8.25.

To use the instrument, place the pointer on the boundary of the shape, trace around the boundary of the shape in a clockwise direction and read off the total roll from the counter as before. We shall show in a moment that the total roll multiplied by the length of AB gives the area of the shape.

To justify why the polar planimeter works we shall use a similar geometrical argument. Let us consider the motion around the curved 'rectangle' RSTU shown in figure 8.28. This is again made up of four sections and just as before we consider the motion during each.

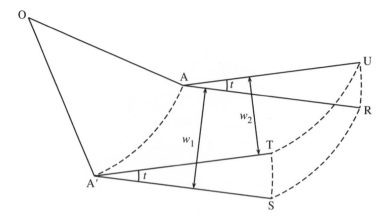

Figure 8.28. Tracing round a curved rectangle RSTU.

1. When moving from R to S the line AB slides and the roll recorded on the wheel will be w_1, the perpendicular component.
2. From S to T the line rotates about the angle t. We do not know what the roll recorded here will be but as before this will cancel out in a moment.
3. Next, from T to U we have a sliding motion and the roll recorded is the perpendicular distance w_2. This is in the opposite direction to w_1.
4. Finally, as the pointer moves from U to R we have the line rotating back in the opposite direction to its original position, i.e. through an angle of $-t$. During this motion the roll cancels out any roll during step 2 above.

Thus, just as before, the total roll recorded will be $w_1 - w_2$, and the area swept out will be this multiplied by the length of the line AB. The line passes over each point of the curvilinear parallelogram exactly once in the positive sense—any other points that it passes over it does so once forwards and once backwards. Hence the area of RSTU is the total roll multiplied by the length of AB. This argument, adapted from Cundy and Rollet (1961), is concluded by noticing that

> [s]ince any area can be supposed divided into elementary curvilinear parallelograms of this type, the result is general.

This is analogous to dividing a shape into rectangles, as we did for the linear planimeter. Interestingly, our argument shows that the radius OA of the circle on which A is constrained to move does not enter into the equations which give us the area of the shape. Hence, we could (on very shaky ground indeed) connect the theory of the polar and linear planimeters by suggesting that a linear planimeter is nothing more than a polar planimeter in which the arm OA is infinitely long. Of course, this requires a huge leap of faith but does provide a conceptual link between the two, seemingly different, devices.

8.6 The Hatchet Planimeter of Prytz

So at last we come to the planimeter that was examined at the beginning of the chapter. This is known as the hatchet planimeter and was invented in 1875 by the Danish cavalry officer and mathematician Holgar Prytz, whose biography is given in Pedersen (1987). The device itself is disarmingly simple, really only consisting of a bent piece of metal, as shown in figure 8.2. As we shall see, the hatchet planimeter does not give the exact area but only an approximation. This is unlike the other planimeters detailed above, which are mathematically exact. Indeed, it is these errors that make this mathematically much more complex and interesting.

In practice if the distance l is sufficiently long compared with the width of the shape being measured and the starting line passes through the centre of mass of the shape, the mathematical error is very small indeed. Although the mathematical error may be reduced in this way, this improvement is more than cancelled out by the very much increased proportional error in the measurement of the displacement.

On the one hand we have mathematically exact but very complicated devices like the polar planimeter, and on the other a simple device that contains a small mathematical error. Superficially, the former would appear to be better, but since all physical devices contain some error in practice, which one results in the smallest *overall practical error* is far from obvious. For example, the linear and polar planimeters rely on the slip–roll interaction of the wheel. Many things can upset this. Friction in the wheel bearings, worm gears, etc., will certainly reduce the actual roll recorded. Corrosion on the wheel surface or the frictional qualities of the paper will introduce other, hard to quantify inaccuracies. Some planimeters solved this latter problem by providing a prepared surface on which the wheel rolls. The wheel and Vernier scale are marked out to allow readings to within 0.001 of a roll, and the whole carefully balanced roller spindle must be mounted exactly parallel to the line AB. Hence the simple hatchet planimeter, with a blade which can be easily sharpened at home, may well be a much more accurate solution. It is certainly the most economical, since Farthing (1985) quotes the starting cost of a polar planimeter as being £165 in 1982, which with inflation is equivalent to £400 in 2006.

The sources of the errors in the hatchet planimeter was a matter of much controversy, which was discussed in the journal *Engineering* during the period 1894-96. The discussion begins with a paper on 'Integrating instruments' by Henrici, which was read at the Physical Society on 23 March 1894. This was followed on 25 May 1894 by an article on p. 687 of *Engineering* giving a geometrical explanation of how the hatchet planimeter is able to measure a plane area. On 22 June 1894 Prytz gave a mathematical explanation of the source of error in the hatchet planimeter and on the same day various planimeters were discussed at another meeting of the Physical Society. This meeting is reported on pp. 26-27 of the 6 July issue of *Engineering*. There seems to have been some confusion as to the source of this error and in the 14 August 1896 issue of the journal (p. 205) Ernest Scott published some 'improvements'. Actually, these amounted to little more than (i) being able to alter the length of the instrument and (ii) adding a wheel to record the final displacement by rolling the wheel back to its starting point, just like the recording wheel of the Amsler planimeter in fact. These modifications are illustrated in Scott's figure 5 in his journal article. One of his other innovations was to replace the pointer with a rotating disc. Scott claims that in its original form the instrument, which he calls a stang planimeter, has 'certain physical defects, which

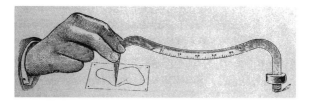

Figure 8.29. Goodman's planimeter.

have probably been to a great extent the cause of the vague and erroneous statements one hears with regard to its theoretical accuracy'. He goes on to say that 'the faults of the original stang planimeter were purely physical, and the writer ventures to think he has overcome these at a very slight extra expense by means of the above described improvements'. Actually, Scott was incorrect in asserting that the faults were physical: there are mathematical reasons why the hatchet planimeter only gives an approximation to the area.

A week later, on 21 August 1896, there appears on pp. 255–56 an article titled 'Goodman's hatchet planimeter'. This claims to be another improvement on the hatchet planimeter that eliminates the sources of the errors. Goodman does eliminate one of the two sources of error in the hatchet planimeter by measuring *arc length* rather than *linear displacement* by means of a curved scale placed on the back of the instrument. This can be seen in figure 8.29. Of course, if the angle is small then the chord is an excellent approximation to the arc length. However, in this article he states that the starting point is chosen 'somewhere near the centre of the figure; the exact position is, however immaterial'. This latter statement is incorrect since one source of error is only eliminated if the centre of mass is the initial point. Two letters appeared on p. 308 of *Engineering* on 4 September 1896 discussing the relative merits of Goodman's and Scott's 'improvements'. However, a week later Prytz entered the fray with a reply. In this he criticizes Scott, saying that he 'goes too far' in his claim that the hatchet planimeter is 'as accurate [as], if not more so than, the Amsler'. More importantly, he reasserts that the hatchet planimeter is not mathematically correct. He discusses the practical problems of using a very long hatchet

Figure 8.30. Prytz's displacement scale.

planimeter to measure small areas and suggests that 'this incon-
venience may be removed by circumscribing the area two or
more times before the end mark is set, as is indicated in the
short explanation that accompanies every instrument'. For large
areas he recommends dividing the area and measuring the parts.
These two simple techniques completely remove the need for a
changeable length planimeter. He also comments on the use of
a wheel rather than the knife blade or keel.

> I must remark that when the edge of the keel is grown dull,
> it is easy for any one to sharpen it a little. A wheel must ever
> be more sensible to dust and rough usage, and the more, the
> sharper and more accurately constructed it is.

Prytz goes on to complain further:

> I wonder why Mr. Scott puts his name to the instrument
> instead of mine. He has on no point changed the principles
> for its use as a planimeter, and all the 'improvements' that
> he proposes are not new.

In an adjacent letter on p. 347 Prytz then rounds on Goodman,
ending his letter with the following plea (see figure 8.30).

> I must council the engineers, rather than use the 'improved
> stang planimeters' to let a country blacksmith make a copy
> of the original instrument, and, if they wish to avoid the mea-
> suring or calculating after a common rule, then, themselves
> to draw on the paper whereupon the keel glides a scale.

A week later on p. 337 of the 18 September edition of *Engineering*
Goodman replies to both Scott and Prytz, 'with the hope of easing
their troubled minds'. His tone is far from conciliatory, and his
letter closes as follows.

I have very good reasons for using the 'bad form' of the tracing points I do, and I think I know what I am about. I happen
to have one of Captain Prytz's planimeters, which I can quite
believe may have been made by a country blacksmith; for my
own part though, I prefer having mine made by a mathematical instrument maker, for curiously enough, I find the latter
usually works more accurately that the former. Possibly in
Denmark things are different.

Practical men have no time to waste in calculating the
mean square of the radii of all the figures they require the
area of. They want an instrument which gives a close approximation to the truth, without any calculation whatever. Such
is mine. Let those who doubt get one and try for themselves.

So perhaps we should now explain why with the aid of the
hatchet planimeter you are able to calculate a good approximation to the area of a plane shape. This will take a similar informal
geometrical approach, using much that has been developed for
the linear and polar planimeters. In doing this we shall point
out the mathematical source of the discrepancy between what
is measured by the hatchet planimeter and the true area of the
shape being traced around. This will explain exactly what Prytz,
Scott and Goodman were arguing about.

8.7   The Return of the Bent Coat-Hanger

To begin our discussion we shall return to our friend the line AB.
In order to explain how the experiment at the beginning of the
chapter is connected with the linear and polar planimeters we
shall consider the line AB moving more generally in the plane.

To be more definite we assume that A and B each move once
around a closed curve and return to their original positions. We
assume further that during this motion the line does not make a
complete revolution. We also assume that the situation is essentially that shown in figure 8.31, where B is always to the right of
A. As before, we fix a wheel to the line using AB as an axle and at
various points consider the roll recorded. To understand motion
such as this we shall consider the moving line in two ways.

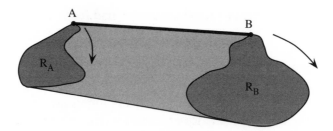

Figure 8.31. The area swept out by a line.

1. First we consider the microscopic, or rather infinitesimal, movement of the line and appeal to calculus.
2. Second we consider the macroscopic case, by application of pure geometry.

By relating these we will justify why the Prytz planimeter works. We sketch an informal argument that appeals to some simple ideas from calculus.

To begin, then, let us take a small motion, shown in figure 8.32, as the line moves from AB to A′B′. During any such very small motion the area swept out can be decomposed into two portions. The first is a rotating motion during which one end is fixed at A and the other moves from B to C. The second is a sliding motion from A to A′ when the other end moves from C to B′. We can approximate the total area swept out as the sum of the two components and, using previous calculations from section 8.3, we arrive at

$$\text{infinitesimal area swept out} = \text{area}(ABCB'A')$$
$$= \text{area}(ACB'A') + \text{area}(ABC)$$
$$= l\,\mathrm{d}w + \frac{1}{2c}l^2\,\mathrm{d}t.$$

What we shall do is take infinitely small movements and sum these to calculate the area swept out during the whole motion shown in figure 8.31. By integrating such infinitesimally small $\mathrm{d}w$ and $\mathrm{d}t$ it is possible to show that the area swept out by the line AB is

$$\text{area swept out} = l\int \mathrm{d}w + \frac{1}{2c}l^2\int \mathrm{d}t. \qquad (8.1)$$

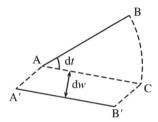

Figure 8.32. A small motion of the line.

Moving A and B simultaneously around R_A and R_B, respectively, and returning them to their original positions, we have assumed that the line AB does not make a full revolution. Furthermore, the total angle sums to zero. Compare this argument—that the total angle is zero—with the symmetry of the rolls w_2 and w_4 in the argument for the linear planimeter. A consequence of no overall rotation is that the total roll recorded is due only to the sliding of the arm AB. Thus, the formula (8.1) reduces to

$$\text{area swept out} = l \int dw.$$

So, if w is the total recorded roll, then the total area swept out by the arm during the entire motion is lw, as before.

Turning to the macroscopic motion of the line we consider which areas of the plane are actually swept out during our motion. This is illustrated in figure 8.31, which you should compare with figure 8.22. The point B traces around the region R_B once. During this motion the line sweeps over the region once in a positive sense. Simultaneously, as the point A traces around R_A, the line sweeps over this region once in a negative sense. What about the region in between? It turns out that this is swept over as many times in the positive sense as in the negative sense so that, during the entire motion, the remaining area is exactly

$$\text{area}(R_B) - \text{area}(R_A). \tag{8.2}$$

This purely geometrical argument can be used in conjunction with the calculus argument to obtain the crucial relationship

$$lw = \text{area swept out} = \text{area}(R_B) - \text{area}(R_A). \tag{8.3}$$

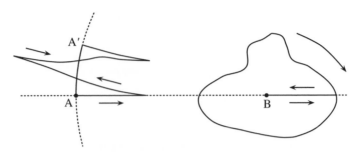

Figure 8.33. A pursuit curve.

Notice that (8.3) is general for a line of fixed length in the plane when the end points move and return to their original positions and the whole line does not perform a complete revolution. Now we relate (8.3) to planimeters.

First we consider how the point A moves in a linear or polar planimeter. For the linear planimeter, A moves along a straight line, while for the polar planimeter, A moves forwards and backwards on an arc of a circle. In both cases the area of R_A, the region enclosed by the motion of A, is zero. Hence, (8.3) reduces to give area(R_B) equal to lw: the total recorded roll multiplied by the length of AB. What about the hatchet planimeter?

To help with the discussion, let us take the region R_B as shown in figure 8.33. The pointer begins its motion at the point B inside the region R_B as near as possible to the centre of mass of R_B. Begin the motion by moving along the straight line on which AB lies and then move around the boundary of the region R_B once in a clockwise direction. Finally, return along the line to the original position B. The key to the hatchet planimeter is the observation that as the pointer traces around the boundary of the region the knife blade follows in a pursuit curve. This curve of pursuit is a zigzag path arriving at A′ when the pointer returns to its original position. Imagine a wheel W fixed at the point A itself, using the line AB as an axle as before. Since the hatchet blade cannot move in a direction perpendicular to AB the total roll recorded along the pursuit curve from A to A′ will be zero.

Now, to use the formula (8.3) we must have a closed curve enclosing R_A. Therefore, to create a closed curve we imagine the hatchet blade being forced back to its original position A along a circular arc centred at B. This violates the non-slip condition. However, when we force the blade back to A, the roll recorded will be the arc length from A′ to A. Now if, as we have assumed, this angle is small, we can approximate the curve length AA′ well by the straight-line distance between A and A′.

We are now in a position to apply our formula (8.3), since A and B have completed their motion and returned to their original positions. Estimating the total 'roll' as the straight-line distance AA′ we have

$$\text{area}(R_B) - \text{area}(R_A) = \text{area swept out} \approx l\text{AA}'.$$

Hence

$$\text{area}(R_B) \approx l\text{AA}' + \text{area}(R_A).$$

Now we see two sources of error:

1. the line AA′ approximates the circular arc AA′, with centre at B;
2. area(R_A) may not be zero.

To make sure that the first of these is small we take the length of AB to be long in relation to the maximum diameter of the shape we are trying to measure.

The area of R_A is a much more difficult problem. For the planimeter to work we must be able to ensure that this is very small. Draw a circle of radius l centred at the centre of mass of B. Part of R_A will be inside this and will have positive oriented area and part of R_A will lie outside this and will have negative area. Indeed, it can be shown that if the starting point of B is the centre of mass, then the area of R_A will be zero! However, finding the centre of mass is just as hard as calculating the area and so in practice one needs to make an educated guess and estimate the position.

The justification given here that each device works is not rigorous mathematics, rather it seeks to give an intuitive geometrical explanation. The polar planimeter can be justified rigorously using some advanced vector calculus and a result known as Green's theorem—material that is often only encountered midway through a university mathematics degree. The details of this theory can be found in, for example, Gatterdam (1981). The error analysis of the hatchet planimeter is examined by Farthing (1985) and Crathorne (1908). That given by Foote (1998) employs even more advanced differential geometry.

8.8   Other Mathematical Integrators

Before the invention of the digital computer ingenious inventors devised mechanical instruments which performed a whole range of other calculations. These included finding the centre of mass, moment of inertia and so on. Some of these were based on the following disc integrator.

Imagine that we have a function $f(x)$ that we wish to integrate. That is to say, we want to calculate $\int_0^t f(x)\,dx$. Imagine a horizontal disc which is spinning at a constant speed: one revolution per second, say. Riding on top of this is a vertical disc which is driven by the horizontal one. The axes of these two discs intersect and the distance of the vertical disc from the centre of the horizontal disc can be varied.

We build an input–output mechanism from this arrangement by taking the distance of the vertical disc from the centre of the horizontal disc as the input. We take t as a time variable and the output $g(t)$ as the total rotation of the vertical disc. Now, the speed of rotation $g'(t)$ will be proportional to the distance of the vertical disc from the centre of the horizontal disc. Since our input, $f(t)$, is this distance, we have

$$g'(t) = kf(t),$$

where k is some constant of proportionality. Using the fundamental theorem of calculus we can solve the above equation to give $g(t)$ as

$$g(t) = g(0) + k \int_0^t f(x)\,dx.$$

The device essentially integrates the input. Of course, there are many subtleties associated with such a device. Errors occur due to the very subtle slip–roll interaction and other physical problems—see Hele Shaw (1885) for more explanation. To do anything useful the force or movement needs to be amplified, perhaps with a configuration of linkages as in chapter 2. However, before the invention of the digital computer, such integrators were being chained together to solve complex differential equations. This is itself an application about which whole books have been written. You could construct models of such devices. In a straightforward survey article on this subject, Hartree (1938) comments as follows.

> My first impression on seeing the photographs of Dr Bush's machine was that they looked as if someone had been enjoying himself with an extra large Meccano set. This suggested the possibility of trying to build... a model version of the machine.... The result was successful beyond expectation; we found it possible to build almost entirely in Meccano, a model differential analyser, which would work not only qualitatively but quantitatively to an accuracy of about 2 per cent.

Indeed, Hartree mentions the construction of such devices as school projects.

Chapter 9

ALL APPROXIMATIONS ARE RATIONAL

It is obvious that irrationals are uninteresting to the engineer since he is only concerned with approximations and all approximations are rational.

Hardy (1967)

Imagine that you have a secret so special, so shocking in its implications, that the members of an organization to which you belong are prepared to kill to keep it. This was exactly the situation in Greece in the fifth century B.C.E., approximately a hundred years *before* Euclid wrote his *Elements*. The organization was the Pythagorean brotherhood, founded around 540 B.C.E. by Pythagoras of Samos (580–500 B.C.E.). They were a monastic organization dedicated to mathematics and number worship who studied numbers and their properties. Their starting point was a view of the world which derived from concrete whole numbers, the kind of thing we are all familiar with from counting. From this it is a natural step to fractions, $\frac{a}{b}$, which are also called rational or *ratio* numbers. It is natural to suppose that all numbers can be expressed as a fraction of some kind if we are prepared to take a and b large enough. The secret you have to keep is that this is actually *false*. That is to say, some numbers cannot be written as ratios, i.e. as rational numbers. Folklore has it that a member of the Pythagorean brotherhood was thrown from a boat and murdered for disclosing the secret that Hardy brushes aside as 'uninteresting to the engineer'.

Hardy is certainly correct, as we have already seen, in highlighting the need for accurate approximations in engineering. Many of these result in ratios of whole numbers and it is examples of these that we shall examine in this chapter. However, Hardy was

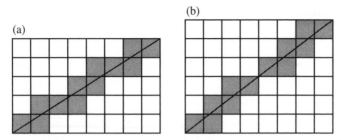

Figure 9.1. Laying pipes under a tiled floor.

quite wrong when he claims that 'irrationals are uninteresting to the engineer'. In section 10.11 the difference between rational and irrational numbers turns out to be crucial. We shall begin this chapter by explaining what rational numbers are and then go on to detail some applications. To introduce and illustrate these ideas we have to resurrect the fiction of the thin line.

9.1   Laying Pipes under a Tiled Floor

Here we return to our fictitious thin line and consider how this lies upon a grid of squares. This has become a popular problem, which is often posed by asking how many tiles one would need to lift in order to place a pipe diagonally underneath a rectangular floor. So let us imagine we have a rectangle of n by m squares. The question simply becomes, how many squares does the diagonal line cross? Two examples of this are given in figure 9.1, with an 8×5 grid on the left and an 8×6 grid on the right.

In both examples the line crosses twelve squares, however, there is a significant difference. The second diagonal line actually passes exactly through the corner of a tile. That is to say, it passes through the intersection of a vertical and a horizontal line. When this happens we could separate the problem into two smaller ones, consisting of the bottom left and top right rectangles. It is not hard to see that when there is a number $k \neq 1$ which divides both n and m exactly, then a line passes through a corner. This is sometimes expressed by saying that n and m share a common factor.

If n and m have no common factor, mathematicians say they are *co-prime*. In this case, the diagonal will leave one square and enter the next either by crossing a horizontal line or by crossing a vertical line. So the number of times it leaves one square and enters the next will be equal to the number of horizontal lines it crosses, i.e. $m - 1$, added to the number of vertical lines it crosses, i.e. $n - 1$. So the number of squares it occupies will be one more than this, since it starts in one square. Hence the number of squares is just $n + m - 1$. But only in the case when n and m share no common factor. Since this problem is so popular, we will not spoil the fun by giving the full solution in the general case, although the ingredients are all here. Details of this, and extensions to the problem, are given in Branfield (1969).

What is more interesting still is to start with an arbitrary line. We make the assumption that the line does pass through the intersection of two grid lines somewhere. We might as well take this part to be the origin of a coordinate system, so we know that the line really does pass through the $(0, 0)$ point. What we need to do is look along the line and find the next point on the grid at which the line passes through a corner. We shall call this point, as before, (n, m). So in the first of the two examples in figure 9.1 the point was $(8, 5)$, and in the second the *first such point* was $(4, 3)$, rather than the next point $(8, 6)$, which is shown.

But does such a point always exist?

The rather surprising answer to this question is a resounding *no!* That is to say, there are lines on the grid which pass through the point $(0, 0)$ and *no other corners*. They continue infinitely in both directions missing all other corners. To visualize this, imagine looking at the grid of squares as you accelerate away in a space craft. More and more of the graph paper comes in to view, and the squares appear to become tiny. Still, no matter how much of the grid you view, the line never again exactly crosses the corner of another square, even if it comes very close indeed to doing so. To explain how this might be, we consider the *gradient* of the line. This is a measure of the slope and is defined as

$$\text{gradient} = \frac{\text{change in vertical}}{\text{change in horizontal}}.$$

For example, in figure 9.1(a) the gradient is calculated as a (horizontal) change m of 8 units, for each (vertical) change n of

5 units. This gives a gradient of $\frac{5}{8}$. In figure 9.1(b) the gradient can be calculated as either $\frac{6}{8}$ or $\frac{3}{4}$. However, these fractions are the same quantity, fortunately giving us a unique numerical value for the gradient when we count squares.

The question of when the line next crosses over a corner of a square is now equivalent to being able to express the gradient, z say, of the line as a fraction:

$$\text{gradient} = z = \frac{m}{n}.$$

That is to say, is it possible to write any number z as a fraction? If so, mathematicians say this is a *ratio* or *rational* number, where n and m are whole numbers. Numbers that cannot be expressed as a fraction of whole numbers are known as *irrational* numbers. This does not mean they lack common sense or behave irrationally, rather that they cannot be written as a ratio. The existence of such irrational numbers came as a great shock to the ancient Greeks, and in particular the Pythagoreans, whose concept of number itself seems to have been based upon the notion of ratios of lengths.

The most famous example of such an irrational number is the length of the diagonal of the unit square, which may be constructed without the slightest hesitation. By the Pythagorean theorem this length is $\sqrt{2}$, and as a number is certainly well defined. However, it turns out not to be possible to write this as a ratio. So, for completeness, let us examine the proof that the square root of 2, that is $\sqrt{2}$, is irrational.

In figure 9.2 we have a unit square, which has a diagonal of length $\sqrt{2}$. Hence, the gradient of the line shown is also $\sqrt{2}$. Let us assume that $\sqrt{2} = m/n$, where m and n are integers and the fraction is written in the lowest terms. This is tantamount to asking for the first time a line with this gradient crosses over a corner. Then multiplying by n and squaring both sides gives $2n^2 = m^2$. That is to say, m^2 is even. Now, from this one may infer that m itself is even, and we write $m = 2p$ for some integer p. Substituting for m we now have $2n^2 = 4p^2$, or $n^2 = 2p^2$. Thus n^2 is even and, by an identical argument, n must be even. Thus

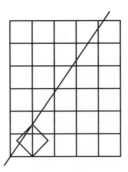

Figure 9.2. A line of gradient $\sqrt{2}$.

$n = 2q$, for some other integer q. Hence we have

$$\sqrt{2} = \frac{m}{n} = \frac{2p}{2q}.$$

But we assumed that m/n was written in the lowest terms, so we have a contradiction. Thus $\sqrt{2}$ must be irrational.

Actually the situation is not hopeless, since although we cannot write an irrational number exactly as a fraction, we may approximate it as closely as we choose using rational numbers. It is exactly this process to which Hardy was referring. That is to say, given any irrational number z, and any error tolerance $\varepsilon > 0$, then there do exist m and n such that

$$\left| z - \frac{m}{n} \right| < \varepsilon.$$

For example, if $z = \pi$, then it can be approximated by $\frac{22}{7}$ and rather better by $\frac{355}{113}$, which satisfies $|\pi - \frac{355}{113}| < 10^{-5}$.

Although we have shown that irrational numbers exist, one might be forgiven for asking what all the fuss is about. After all, we can approximate any number as closely as we would wish using rational numbers. One way to approximate π more closely is to write part of the decimal expansion, and form a fraction, such as $\frac{31\,415\,927}{10\,000\,000}$. Conversely, as Hardy points out in the quote at the beginning of the chapter, any such approximation will be a rational number. If you are trying to measure a length, does this distinction matter? Why does the mathematician or engineer care about the difference between a rational and an irrational number?

One reason is illustrated by the A series of standard international paper sizes. There are two considerations affecting the shape of a sheet of paper. Its proportions must be pleasant to look at and handle, and these proportions should be preserved when the sheet is folded in half. These criteria are satisfied if the proportions are $1:\sqrt{2}$. While it is possible to prove this is true, it is less helpful when actually trying to make sheets of paper. What is the value of $\sqrt{2}$? Is it good enough to use 1.4 as an approximation or would it be better to take 1.41 or 1.414? It is better for machine settings if $\sqrt{2}$ is expressed as a rational approximation, and better still if this ratio is internationally accepted.

Beiler (1964) describes how $\sqrt{2}$, or indeed the square root of any integer, can be approximated by a series of increasing accuracy. Without going into details of the calculations the series for $\sqrt{2}$ is

$$\frac{1}{1}, \quad \frac{3}{2}, \quad \frac{7}{5}, \quad \frac{17}{12}, \quad \frac{41}{29}, \quad \frac{99}{70}, \quad \frac{239}{169}, \quad \dots$$

You will notice that if we multiply the top and bottom of the sixth fraction, or convergent, by three we obtain the size of a sheet of A4 paper in millimetres. The A series of international paper sizes is given in Chambers dictionary, for example, and a few common sizes are shown here together with the square of the ratios of their sizes:

A3	420 mm × 297 mm	1.9998
A4	297 mm × 210 mm	2.0002
A5	210 mm × 148 mm	2.0133
A6	148 mm × 105 mm	1.9867

Sizes are increased by doubling the length of the shorter side and decreased by halving the longer side. Similar series, B and C, are used for posters and envelopes.

9.2   Cogs and Millwrights

Suppose that we wish to connect two shafts together with gears so that one runs at twice the speed of the other. Suppose further that the exact ratio of two to one may not be critical and some latitude around it is allowed. One way of doing this is to have a 20-toothed gear meshing with one of 40 teeth, giving a ratio of exactly two to one. This implies that a tooth on the larger wheel will always mesh with one particular tooth on the smaller. Looking at it from the point of view of the smaller wheel, any one of its teeth will always mesh with the same two diametrically opposite teeth on the larger wheel. This happens at every turn of the 40-toothed gear, and for every two turns of the smaller one. This point has been laboured so that the problem caused by one of the gears having an imperfect or badly formed tooth can be readily appreciated. The extra wear or damage will be confined to one or two teeth on the other gear and there is the possibility of jerky or erratic transmission.

In order to overcome this localized wear an extra tooth could be added, a *hunting cog*, so that a ratio of 20:41 could be used instead. By this means wear will be distributed evenly among all teeth as only after 41 turns of the small wheel will the same teeth be meshing again. As it happens 41 is a prime number, but what is crucial is that the number of teeth should be prime to each other, as for example 21:38, and not that the number of teeth on any gear is necessarily prime. This even wear has been gained at the expense of losing the exact ratio of 2:1. The early millwrights, using the wooden gearing with inserted teeth used in wind- or water-driven machinery, recognized the advantage of this as they rarely needed an exact ratio of, say, 2:1.

Exact ratios are possible where a train of gears is used, and to illustrate this we need to make two quite realistic and practical assumptions: the minimum number of teeth is 20, and the gear ratio should not exceed 6:1. We take as an example a required exact ratio of 360:1 and first note that

$$6^4 > 360 > 6^3.$$

This means we shall need a train of four shafts and a total of eight gear wheels. Factoring, we have

$$360 = 2 \times 2 \times 2 \times 3 \times 3 \times 5,$$

and since we need four pairs of meshing gears we write this as

$$\frac{360}{1} = 2 \times 2 \times 2 \times 3 \times 3 \times 5 \times \frac{20 \times 20 \times 20 \times 20}{20 \times 20 \times 20 \times 20}.$$

Note that 20 is the agreed minimum number of teeth. The train could then be

$$\frac{100}{20}, \quad \frac{100}{20}, \quad \frac{90}{20}, \quad \frac{64}{20},$$

or

$$\frac{100}{20}, \quad \frac{96}{20}, \quad \frac{80}{20}, \quad \frac{25}{20}.$$

If we now require each pair of gears to be relatively prime, we proceed in a slightly different manner and write

$$\frac{360}{1} = 2 \times 2 \times 2 \times 3 \times 3 \times 5 \times \frac{20 \times 21 \times 22 \times 23}{20 \times 21 \times 22 \times 23},$$

giving a gear train

$$\frac{99}{20}, \quad \frac{92}{21}, \quad \frac{105}{22}, \quad \frac{80}{23}.$$

The overall ratio of 360:1 has been preserved and the gears are prime to each other, but we have the expense of needing eight gears all of different sizes, compared with only four or five in the design of the first train.

So much for the arithmetic from Schwamb and Merrill (1984), it is now time to hear the views of two very different experts as to the use of hunting cogs. We start with Schwamb again.

> In cast gears, which will be more or less imperfect, it would be much better if any imperfection on any tooth could distribute its effect over many teeth rather than that all the wear due to such imperfection should come always upon the same tooth.

Unsurprisingly, a manufacturer of cast gears took a different view.

> The object is to secure even wearing action; each tooth will have to work with many other teeth, and the supposition is that all the teeth will eventually and mysteriously be worn to some indefinite but true shape.
>
> It would seem to be the better practice to have each tooth work with as few teeth as possible, for if it is out of shape it will damage all teeth that it works with, and the damage should be confined within as narrow limits as possible. If a bad tooth works with a good one it will ruin it, and if it works with a dozen it will ruin all of them. It is the better plan to have all the teeth as near as perfect as possible, and to correct all evident imperfections as soon as discovered.
>
> Grant (1907, section 46)

These opinions are some 100 years old, and since then manufacturing techniques have improved considerably and allied with advances in metrology and inspection procedures, there is little need to use hunting cogs in current engineering practice. Nevertheless, these cogs were the fruits of applying mathematics to what seemed to be a real engineering problem.

9.3 Cutting a Metric Screw

We would like to widen the discussion of gear ratios to include a problem faced by those model engineers whose lathes have imperial graduations but who need to work to metric standards. There are no problems if a piece of work has to be turned to a particular diameter, or parted off to a particular length, because it is easy enough to convert millimetre measurements to imperial with a pocket calculator. The problem is cutting metric threads on an old imperial lathe. In order to cut a thread it is necessary for the cutting head to be moved along the length of the lathe a distance exactly equal to the pitch of the thread for each revolution of the workpiece.

This movement is achieved by coupling the lathe spindle, which holds the work, to the carriage that holds the cutting tool.

This coupling is achieved by a long screw, known as the lead screw, and is provided by a train of gear wheels, called change wheels, that can be set to allow threads of different pitches to be cut. Many imperial standard lathes have lead screws of 8 tpi (threads per inch) so that one turn moves the tool 0.125 in, and a thread of this pitch can be cut if it turns at the same speed as the work, a ratio of 1:1. To cut a thread of say 10 tpi needs a train that gives the screw a reduction of 4:5.

Metric threads are based on multiples of 1 mm pitch, so, remembering that 1 in equals 25.4 mm, a ratio of $1:\frac{25.4}{8}$, or 1:3.175, is needed. A useful way of working this ratio is 8:25.4 or 40:127. Apart from 2, 127 is the only factor of 254. The fact that it is prime is not unduly worrying. The difficulty is that a gear with a large number of teeth is often inconveniently large to fit into a train designed for smaller gears, and may not even fit inside the gearbox housing. Therefore we seek a ratio of smaller numbers that cannot be exact, obviously, but which gives a sufficiently good approximation. A surprisingly good one is

$$\frac{65}{55} = \frac{150.0909}{127} \approx \frac{150}{127}.$$

This is the coupling required for any metric thread as whatever the pitch the factor of 127 is implicit. To cut a thread of 1 mm pitch a train

$$\frac{65}{30}\frac{20}{50}\frac{20}{55}$$

is used. These fractions should be read as 65 teeth (T) on the work spindle driving a 30 T wheel, coupled to a 20 T wheel, driving a 50 T wheel, ending with the 55 T wheel on the screw moving the tool along the bed of the lathe. The result is a thread of 1.000 606 mm pitch, and although obviously not suitable for making a micrometer or other high-precision piece of equipment it is perfectly acceptable for amateur model work where the length of thread is generally quite limited. To put the approximation into perspective a human hair is about 0.04 mm in diameter.

In purely mathematical terms the ratios in the gear train could have been written as

$$\frac{20}{30}\frac{20}{50}\frac{65}{55} \quad \text{or} \quad \frac{20}{30}\frac{65}{50}\frac{20}{55},$$

since the order of multiplication and division does not matter. These alternative ways of working the ratios cannot be formed into a gear train, however, because of interference of gears and the stub shafts to which they are attached.

9.4 The Binary Calendar

Another very familiar situation in which irrational numbers cannot be ignored is the calendar. Here we are concerned with the length of a year, defined for our purposes as the period of time between consecutive spring equinoxes. This particular definition of a year is also known as the tropical year.

It is unfortunate for us that this period of time does not contain an exact number of days. In fact, the length of the year 2000 was about 365.242 199 days long. What then is a day? We shall take a day to be the period of time between consecutive moments at which the Sun is at its highest point in the sky, which is often called a solar day. This definition is most useful when constructing a sundial, since this moment is relatively easy to pin down, unlike midnight for example. We shall assume that the length of a year in days remains constant, although in fact the speed of rotation of the Earth on its axis is slowing, but by only half a second or so per century.

For obvious but purely practical reasons it is very convenient to fit a whole number of days into a year. If we fixed a year as 365 days, then quite quickly we would find the months drifting apart from the seasons with which we associate them. So December would eventually move to mid summer. People are not comfortable with this idea, and as a result extra days are added to keep the seasons in step with the calendar. These are the leap years with which we are all familiar, in which an extra day is added to the end of February.

You will notice that the length of the year is close to $365\frac{1}{4} =$ 365.25 days. And indeed one of the first attempts to improve the calendar occurred in 46 B.C.E. when the Julian calendar was enacted. This adds an extra day to every fourth year, giving the average length of a year as 365.25 days. This is not quite correct. If the length of the year was a rational number of days then it would be possible to correct this drift accurately. Unfortunately this is not the case, and the Julian calendar adds about 0.007 801 of a day extra per year. This might not seem like much, but it amounts to about 11 minutes. Over time, the Julian calendar adds an extra day about once every 128 years.

This discrepancy was noticed by Roger Bacon in 1267, who wrote to Pope Clement IV that

> [t]he calendar is intolerable to all wisdom, the horror of all astronomy, and a laughing stock from a mathematician's point of view.

Although 128 years is a long time, by the time Roger Bacon noticed this anomaly the calendar had already drifted some nine days and so religious festivals, such as Easter, were being celebrated at the wrong time. However, making such an accusation amounted to heresy in the thirteenth century, for which the penalties were severe, so it must have been with some trepidation that Roger Bacon addressed the pope.

So controversial were Bacon's ideas that it was not until 1582 that the next attempt to fix the calendar was enacted by Pope Gregory XIII. In this Gregorian calendar, which we still use today, the century years, such as 1700, are *not* leap years unless they are multiples of 400. Thus, the year 2000 was a leap year, despite being a century year, and has the honour of being the first year in which the 400-year rule was applied.

In the Julian calendar days are added, but with the Gregorian system leap years are dropped. Even worse, in order to realign the calendar, days were dropped from it. So people who went to bed on 4 October 1582 woke up the next day on what was, according to the new calendar, 15 October. This caused an uproar and few European countries adopted the new system immediately, with Great Britain waiting until 1752. Russia accepted the Gregorian calendar in 1917 and China only in 1949. The Eastern Orthodox Church has still not adopted it. But how accurate is this calendar anyway?

The Gregorian scheme adds 24 leap years for three centuries, then 25 leap years in the next, giving a total of 97 extra days in 400 years. From this it is easy to calculate that the average year length is 365.2425 days. This differs from the true period by 0.000 301 days per year, which means it will be approximately $\frac{1}{0.000\,301} = 3322$ years until a full day's discrepancy has been accrued. Starting from 1582 this gives the first year as about 4904.

We would like to make an alternative suggestion for calendar reform, inspired by the error of one day per 128 years in the Julian scheme and the binary system for writing numbers. Simply write the year (after 2000 say) as a binary number. If the least significant two digits are both zero, then the number is divisible by four and so we have a leap year, unless the first seven digits are also all zero. This means the number is divisible by $2^7 = 128$. Such a test is much simpler than the Gregorian scheme since we have only two rules to apply and not three. Furthermore, the full calendar cycle is only 128 years rather than 400. Even better, most modern computers store numbers in binary format so implementing the rule is extremely straightforward to do. But is it more accurate? This scheme will result in 31 extra days every 128 years, giving an average year length of 365.242 187 days. The difference from the true value is so small that we would have one day too many only after 86 956 years or so. This scheme, as simple as it is, is unlikely to be adopted in the near future.

9.5 The Harmonograph

A rather beautiful way of seeing the importance of rational and irrational numbers is via a harmonograph. In the simplest form of the device two oscillations at right angles to each other are combined. Take two pendulums, for example, and use the position of one to generate the x-coordinate and the position of the

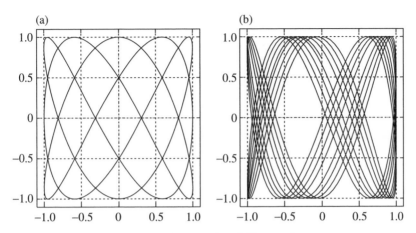

Figure 9.3. Lissajous's figures for different frequency ratios: (a) rational and (b) irrational.

other to generate the y-coordinate. The pictures generated are known as Lissajous's figures.

Let us assume that one frequency oscillates along the x-axis and the other along the y-axis, and that both are sinusoidal oscillations. We assume that the frequency of the horizontal oscillations is a and that along the y-axis is b. Then the ratio a/b will be particularly interesting to us. If a/b is a rational number then eventually we will have a closed curve, such as that shown in figure 9.3(a). However, if a/b is an irrational number then a whole number of x-axis oscillations will *never* equal a whole number of y-axis oscillations and so the curve will never arrive back in its original position. We assume that both oscillations have the same amplitude of one unit. Since we are probably going to need to use broad lines to represent the path of the point, the whole unit square will eventually be filled by our broad line. In fact, no matter how thin we take our broad line to be we will still fill the unit square, which is to say that the path of the point will pass arbitrarily close to every point in the unit square. A small portion of such a curve is shown in figure 9.3(b).

Drawing such figures makes a fun experiment. On a computer one only needs to plot $(\cos(at), \sin(bt))$ for different values of a and b. You can think of the t variable as time. It is also possible to make a simple device that combines the motion of two actual

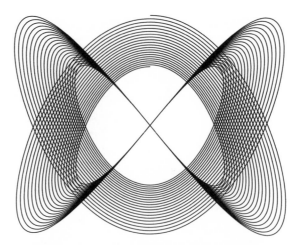

Figure 9.4. A damped Lissajous figure.

pendulums and will draw the curves for you. While the theoretical results in figure 9.3 are generated from perpetual motion, in practice there will be some air resistance and the drag of the pen on the paper to dampen the motion of the pendulums. Hence there is no hope of a perfect closed curve, even if you manage to make pendulums with a perfect ratio of lengths. Fortunately, rather beautiful patterns result, with spirals, cusps and all sorts of other shapes, such as that shown in figure 9.4.

There are various forms for these, which are known collectively as harmonographs. In these a rigid pendulum is suspended on gimbals and is free to move as a conical pendulum. Both the pen and paper move independently under the influence of such a pendulum.

Plans and practical suggestions for construction can be found in the rather lovely little book of Ashton (1999) and also in Cundy and Rollet (1961) so we will not repeat any here. It is well worth spending the time and trouble to make a harmonograph of your own, although quite a bit of careful work is required. The Victorians used these as a popular entertainment, and a huge number of variations are possible. Be warned, though, that watching a harmonograph can be quite hypnotic as the shapes swing into existence, and the whole game can become quite addictive.

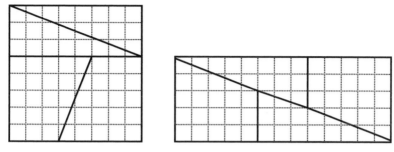

Figure 9.5. Where does the extra apparent area come from?

9.6 A Little Nonsense!

> A heavy warning used to be given that pictures are not rigorous; this has never had its bluff called and has permanently frightened its victims.
>
> Littlewood (1986, p. 54)

We end this chapter with a little more nonsense, which contradicts to some extent the bold quote given above and leaves you with a puzzle. Although this is quite well known, it is an excellent reason why pictures and diagrams *on their own* should be treated with suspicion. A clue to the puzzle can be obtained by considering why it has been placed in this chapter!

We begin with the dissection of a square shown in figure 9.5, which has area $8 \times 8 = 64$. Since this is a relatively simple dissection we recommend you make it for yourself. If you move the four pieces around, it is possible to form the rectangle on the right with area $13 \times 5 = 65$. Where does the apparent extra area come from?

Chapter 10

HOW ROUND IS YOUR CIRCLE?

Round and round the rugged rock the ragged rascal ran.

Nursery rhyme

Imagine you wish to make a circular or spherical object but cannot simply use compasses to mark one out. Alternatively, you might have already made something round and wish to check the accuracy of your construction. We all know what a circle is, so here we discuss the problem of determining if what is supposed to be circular is really circular, and within what limits. This matters a lot in engineering applications of all sorts but particularly with rotating shafts and their bearings.

So let us begin with an experiment for which you need a United Kingdom 50p (or 20p) coin, a 2p coin and, if possible, a new round beer mat or coaster. You will also need the means to measure their widths—Vernier callipers are ideal, but you can just about manage with a clearly graduated rule. Merely glancing at these objects is enough to see that the 2p coin and the beer mat are round and that the 50p piece is certainly not. The next step is to measure their widths, or diameters, in as many orientations as possible. Both coins had constant width, whereas our beer mat had a width varying between 106.1 and 106.5 mm. What can we deduce from these observations and measurements?

The 50p piece as shown in figure 10.1 is obviously not circular, and yet it does have constant width. The beer mat, although appearing to be circular, cannot be because the width varies, and nothing can be said of the 2p coin at this stage except that it might be circular but we do not know for sure. We dismissed the 50p as not round, but what if it had had 77 or 777 sides, still with constant width? It might then have seemed round, but

Figure 10.1. United Kingdom coins: 2p, 20p and 50p.

Figure 10.2. Constructing Reuleaux's rotor.

all we can say is that constant width is a necessary condition for roundness but by itself it is not sufficient. The 2p might be round, but without further tests we cannot be sure. Incidentally, the fact that a 50p piece has constant width is a useful geometric property which helps reduce the number of times such coins get stuck in slot machines.

If you do not have access to these coins, you can make three planar shapes yourself and conduct another experiment. The first is a circle, the second is a square and the third is a slightly unusual shape which can be made easily using the technique that follows—and it is really only a minor variation of figure 5.3. Start with a pair of compasses open and draw a circle. Mark an arbitrary point on the circle and using this as a centre draw another circle. Then draw further circles, all with the same radius, where previously drawn circles intersect (see figure 10.2).

If you focus your attention on the circular arcs, you will notice the shape we have highlighted in grey. There are, of course, six overlapping copies of this shape in the circle. Alternatively, you can construct this shape from an equilateral triangle by drawing

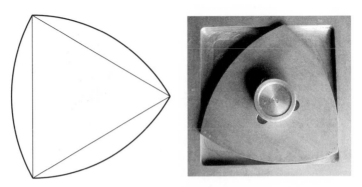

Figure 10.3. Reuleaux's rotor.

circular arcs centred at each vertex passing through the other two. It is known as *Reuleaux's rotor*, after Franz Reuleaux (1829–1905), and is shown as both a schematic and a model in figure 10.3. Historically, the first author to mention such shapes of which we are aware is Euler (1778).

You probably do not actually need to perform the experiment to determine that the width of the circle is constant. Geometrical intuition, together with the Pythagorean theorem, should help you to establish easily that the width of the square varies between 1 and $\sqrt{2}$ times the length of the side. What about Reuleaux's rotor?

It is easy to see that Reuleaux's rotor also has constant width. A tangent to one of the circular arcs will be at a constant distance from a parallel line through the centre of this arc.

This experiment sets the scene for this chapter, where first we discuss the infinite family of closed convex curves of constant width, some of their applications, and how to make examples. We then move on to the more difficult question of how roundness can be measured and why it matters.

A circular cross-section is the most frequently used basic shape in engineering. We define a shape to be *round* if it has a circular cross-section. This is only the case if all points on the boundary are equidistant from the centre, or axis. When measuring the width of the shape we did not make use of an axis, and indeed non-circular shapes of constant width do not possess one. While they can be used as rollers they cannot become true wheels.

Engineers are most concerned with measuring *departure from roundness.* This is a particularly important engineering problem, since many devices depend on rotation. The most important device is the bearing that allows a shaft to rotate smoothly. In a plain bearing a shaft rotates and it is vital to obtain the correct clearance to allow smooth running in the presence of lubrication. More complex types of bearing, such as ball bearings or roller bearings, also depend critically on the roundness of their components.

While the straightness or flatness of an object can be checked against a reference surface, as detailed in section 1.5, this is much more difficult for roundness. In manufacturing it is probably true to say that departures from true roundness present the greatest difficulties of all the form errors that have to be evaluated. In this chapter we shall examine shapes of constant width in two and three dimensions, provide some practical applications of these shapes beyond coin design, and then explain how departure from roundness can be measured.

10.1   Families of Shapes of Constant Width

After the circle the simplest curve of constant width is Reuleaux's rotor, sometimes called Reuleaux's 'triangle'. We shall begin with a variety of families of two-dimensional shapes of constant width. In each of the families there is a spectrum of variation with Reuleaux's rotor at one extreme and the circle at the other. Either the number of sides in a polygon increases or there is continuous change of an arc's radius or construction angle to create this spectrum. Furthermore, all these methods rely on the use of circular arcs to make up the shapes.

The first method is a simple generalization of that used to create a Reuleaux rotor. Start with any regular polygon with an odd number of sides. Draw a circular arc centred at one vertex that passes through the two opposite vertices. The Reuleaux rotor is one example, starting with the regular equilateral triangle. As the

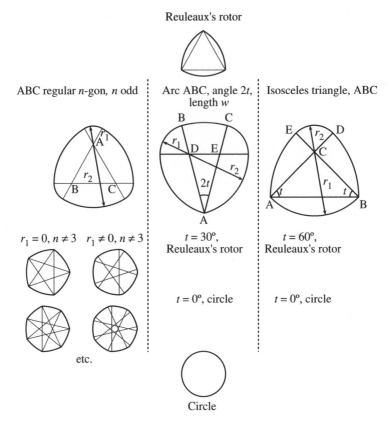

Figure 10.4. Families of shapes of constant width.

number of sides increases so the shapes come closer to true circles, although all these shapes have sharp corners. It is relatively easy to extend the lines from the regular polygons and draw two circular arcs to create a more general form of the Reuleaux rotor. Of course, this can also be done with all the regular polygons. The resulting shapes are now smooth, and all have constant width. This is shown in the left column of figure 10.4

A different family of shapes can be developed, as shown in the middle column of figure 10.4. This family includes a sector of a circle of radius w with angle $2t$. The positions of D and E are determined by simple geometry such that $r_1 + r_2 = w$ with

$$r_1 = \text{BD} = \text{CE} = \frac{w(1 - 2\sin(t))}{2(1 - \sin(t))}, \qquad r_2 = \frac{w}{2(1 - \sin(t))}.$$

Figure 10.5. A collection of shapes of constant width by
G. J. H. Cordle as a final year undergraduate project in 1995.

If $t = 0°$ then the resulting shape is a circle of diameter w, and
if $t = 30°$ then we have a Reuleaux rotor.

The final family is based on an isosceles triangle and is shown
in the right column of figure 10.4. We take A and B to be a distance w apart and draw the isosceles triangle ABC with angles t
at A and B. Hence, we have that

$$r_1 = \text{AC} = \text{BC} = \frac{w}{2\cos(t)}, \qquad r_2 = w - \frac{w}{2\cos(t)}.$$

This shape is unusual in that it is composed of only four circular
arcs. The angle $t = 45°$ is a special case and forms the basis
of a drill that creates a square hole, which we shall describe in
section 10.4.

10.2 Other Shapes of Constant Width

Just in case you harbour any lingering doubt that shapes of constant width, such as the Reuleaux rotor, are a 'special trick', we
shall provide yet more examples. These need not have any symmetries. Let us start with any three lines that form a triangle ABC
as shown in figure 10.6. Choose another point D on the line AB,
external to the triangle, and draw the arc centred at A through

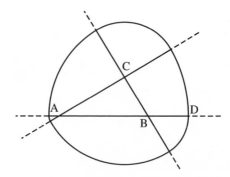

Figure 10.6. Building a shape of constant
width from an arbitrary triangle ABC.

D between the lines AB and AC. Then proceed round the circle
connecting each arc with the preceding one, centring the arc at
the intersection of the lines that it connects. There is a feasibility
constraint: for example, if D is too close to A then the arc centred
at B will fall between A and C. This constraint can always be
satisfied by making the distance AD sufficiently large, and if so
the resulting shape has constant width. This procedure may be
done with an arbitrary number of mutually intersecting lines,
and it results in shapes with an arbitrary number of different
circular arcs.

It is worth making some general comments about all the
shapes so far constructed. It can readily be verified by adding
the lengths of circular arcs in all these examples that the length
of the perimeter of the shape is $w\pi$. This, of course, is identical
to the circumference of the circle with diameter (i.e. width) w.
In fact, a theorem of Barbier, proved rigorously in Lyusternik
(1964), demonstrates that all curves of constant width w have
an identical perimeter length of $w\pi$.

The Reuleaux rotor is in many senses the most extreme shape
of constant width. It has the smallest area for a given width
w, which equals $\frac{1}{2}w^2(\pi - \sqrt{3})$. It also has the sharpest corners, where tangents meet at an angle of 120°. The circle has
the largest area for a given width and all other constant-width
shapes must fall somewhere in between.

All the curves constructed so far have been built piecewise
from circular arcs, but this is not a necessary condition. Now we

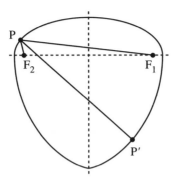

Figure 10.7. A curve of constant width, based on half an ellipse.

turn our attention to even more general methods for creating shapes of constant width that do not rely on circular arcs. Start with a square of width w and draw a convex curve from top to bottom that touches the left-hand side and that is tangent to the top and bottom of the square. At no point should the curvature be less than the curvature of a circle with radius w. We can, in a sense, 'complete' the curve to create a shape of constant width by taking a normal to the curve of length w. This can be done by hand with a ruler very quickly, and reasonably accurately.

As an example of this method we shall create a curve based on one-half of an ellipse:

$$\frac{x^2}{a^2} + \frac{y^2}{b^2} = 1.$$

To start we take only the half with $y > 0$, as shown in figure 10.7. At a point P construct the inward-facing normal PP' of length w. The bottom half is given by the locus of P' as P moves around the top. Figure 10.7 shows the case when $b = \frac{1}{2}a = \frac{1}{4}w$, its minimum possible value. We shall deal more fully with ellipses and their various properties, and explain how to construct the normal to an ellipse, in chapter 13.

There is a limit to the eccentricity of the ellipse used in this construction. The radius of curvature at $(0, b)$ is a^2/b, and the maximum value this can take is $2a$, the length of the major axis, otherwise a convex curve of constant width cannot be constructed. As b increases towards a the whole curve approaches

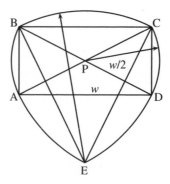

Figure 10.8. Generating a three-dimensional
shape with a spherical cross-section.

a circle. The limit of a Reuleaux rotor cannot be reached with
this particular construction. In section 10.9 we shall provide
another method that has the advantage of providing mathematical expressions for the shapes produced and with which we can
perform some quantitative calculations of the departures from
roundness.

10.3   Three-Dimensional Shapes of Constant Width

A solid may also have constant width. The simplest of these is
formed by rotating the Reuleaux rotor about an axis of symmetry. Indeed, any of the shapes of constant width in figure 10.4
can be rotated about their axis of symmetry, and examples of a
variety of the resulting solids are shown in plate 17.

For another interesting shape of constant width we begin with
a shape based on a right-angled triangle with a hypotenuse of
length w and with the other sides in the length ratio of 2:1. Six
of these triangles overlap and are used to build the skeleton
of a curve of constant width, which is sketched in figure 10.8.
The solid is formed by rotating the shape about a vertical axis
through E. This two-dimensional shape of constant width is
included here because the solid of revolution has the interesting
property that between the planes through AD and BC the surface
is that of the sphere of the same width.

From a practical point of view, the axial symmetric solids
that have circular arcs are by far the easiest to make on a CNC

machine. Five different solids are shown in plate 18. The first paper, of which the authors are aware, that deals with solids of constant width is Minkowski (1904). The topic of solids of constant width is covered in Cadwell (1966), and more fully in Gray (1972). These sources both discuss more general solids that do not have an axis of rotational symmetry. For example, it is possible to begin with a regular tetrahedron and supplement this with four spherical caps of radii equal to the side lengths, which are subsequently rounded off. An example is shown in plate 19. Such solids are significantly harder to manufacture and producing them has only become possible recently with the advance of rapid prototyping techniques.

10.4 Applications

In the United Kingdom, 20p and 50p coins are the most familiar examples of curves of constant width. Here we describe three quite different engineering applications of these curves: as cams, in drills which produce a square hole and pistons in the rotary internal combustion engine. The Reuleaux rotor features in all of these, and we start with cams.

Let us begin with a shape of constant width w, based upon circular arcs. We assume that this is pivoted about one of its vertices and constrained within a yoke of the same width. As it rotates it causes a yoke to move—from left to right, say. Two examples are shown in figure 10.9. Supposing the first of these to be turning clockwise, starting from the position shown, the first $120°$ will move the yoke a distance w to the right. During the next $60°$ the yoke is not moved and this is known as a *dwell*. Then, during the next $120°$, the yoke is pushed back to the left and then dwells for another $60°$ rotation of the cam before repeating the cycle. The important feature of this cam, and any other in the form of a curve of constant width contained within a yoke, is that the cam exerts a positive driving force on the yoke at all stages, and no weight or spring is required at any part of the stroke. Such cams have found applications in the valve control mechanisms of steam engines, such as that of the 1838 engine

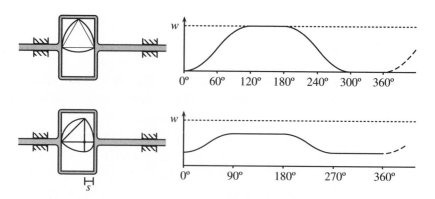

Figure 10.9. Shapes of constant width applied to cam design.

of J. and E. Hall, and also in the feed mechanism for cine film projectors, for example.

An alternative curve can be used that provides two dwells whose periods can be chosen to suit the application. It is based on the isosceles triangle shown in the right-hand column of figure 10.4 and has two dwells of $(180 - 2t)°$ per revolution. This cam, with $t = 45°$, is shown in figure 10.9. Notice that while the displacement between dwells is less, the overall time spent in a dwell position amounts to 180°, i.e. half of the rotation. The actual displacement can be amplified by linkages of course. In practice, the actual cam shape of both curves would allow for the vertices to be rounded to reduce wear in the way shown in the first family of curves in figure 10.4.

A limiting form of cam operates in the 'Scotch yoke', or crank and slotted head, shown in figure 10.10(a). The curve here is a circle pivoted about the axis outside its perimeter, it is a true crank, and not a cam. The design lends itself to a compact form of engine suitable for use in steam pumps for boiler water feed. One end is the piston rod of the steam cylinder and the other is a piston rod from the pump, and the purpose of the crank is to drive a flywheel so that steam can be used expansively and so more economically. A similar yoke device was commonly used for foot wheels to drive watchmaker's lathes before small electric motors were available. They are quiet, of variable speed and can

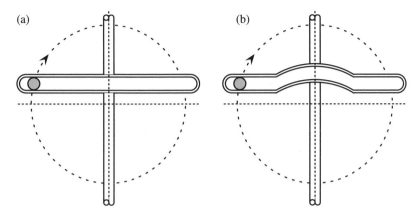

Figure 10.10. (a) The Scotch yoke and (b) the Scotch yoke with dwell.

be reversed at will. We have found them easy to use but have no experience of using such a lathe for long periods of time.

The Scotch yoke is effectively a connecting rod of zero length with the result that as the crank turns it produces simple harmonic motion in the yoke. The same motion can be achieved with a connecting rod only if it is infinitely long. A modified yoke is shown in figure 10.10(b) that provides one dwell per revolution when the radius arc of the yoke equals that of the crank. In a sense this behaves like an inverted cam. The form of rectilinear motion is usually determined by the shape of the rotating cam, but here this motion is governed by the shape of the yoke.

An entirely different use for curves of constant width is as drills or milling cutters to form square holes. One design, which is easily identifiable as a square drill, was proposed by Harry James Watts in 1914. This employs a Reuleaux rotor in which three parts have been removed to provide the cutting edges and to allow the material removed to escape. A model is shown, together with the hole which it cuts, in plate 20. You will notice that it does not remove a perfect square and that small arcs remain in the corners.

The problem of rounded corners is overcome by again using the curve of constant width based on the isosceles triangle, shown on the right of figure 10.4. In particular we use the case with $t = 45°$. If a cam of this shape is housed in a square hole

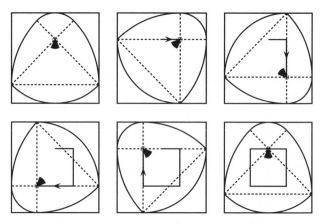

Figure 10.11. Cutting a square hole.

Figure 10.12. Details of a practical cutting tool for a square drill.

of side length w, then part of the quadrant from D to E must always touch one of the sides. Hence, the locus of the point C must follow a square path as the shape is rotated inside a square. A simple calculation reveals that the side length of this is $w(\sqrt{2} - 1)$. In figure 10.11 we sketch the path of C for various orientations of the shape.

A real application of this curve was described in *Mechanical World* in 1939. The need was for a drill to cut blind holes, half an inch square and three-eighths of inch deep, in yellow brass. The cutting tool was a large equilateral triangle cutter.

An alternative form of cutter suitable for demonstration purposes only is shown in figure 10.12. A bar of round tool steel is ground off at one end to give a cutting edge on its centre line. This is held in a bore through the cam, centred at the intersection of the diagonals. The interest arises if the tool is turned through 180° so that the cutting edge faces inwards. Then the result is not a hole but a 'square land' of exactly the same size: of little

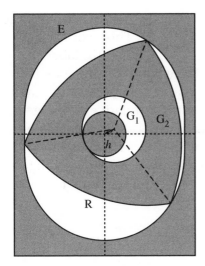 

Figure 10.13. An application to engine design.

practical use, perhaps, but it is fascinating to watch it being made. This drill is shown in plate 21, although unfortunately in this photograph the end of the cutting tool has sheared off during overzealous use.

To use either drill in practice requires a pilot hole to be drilled first so that there is the minimum of material left to receive the final cuts. There is also the question of how to drive each drill, since the axis of any shaft cannot be fixed. The usual solution is to use a device known as an Oldham coupling. The flange at the end of the driving shaft, a fixed axis, is provided with a slot, as is the floating shaft connected to the cam. In between there is a disc with a bar either side, at right angles to the disc, that can slide in the slots. Incidentally, if the axis of the floating shaft is also fixed, a cutting tool in the plane of these axes will cut the disc to a true ellipse. This is the basis for the eccentric chuck frequently used in ornamental turning.

The final application of the Reuleaux rotor is in the design of the rotary-engine car. In a traditional engine the piston recipro-cates but in a rotary engine, as the name implies, direct circular motion is generated. While full details of the engine itself can be found in Cole (1972), for example, for us the most interesting part is the geometry of the engine. With reference to figure 10.13

we take a Reuleaux rotor R and, from the centre, drill a large hole to create an internal gear of radius $3h$. This gear is labelled G_2. This then rotates on a fixed gear of radius $2h$, labelled G_1. If the distance from the centre of the rotor to the apex is $3r$, then the width of the rotor is $6r\cos(30°)$. Physical constraints determine the size of the Reuleaux rotor, relative to the gears. For example, it is necessary that the gears actually lie inside the Reuleaux rotor and this will occur if $h < \frac{3}{5}r$. By symmetry it is clear that the three vertices of the rotor all move along the same path. Taking $R := 2r$ we can show that the path, E, is given by

$$x = (r + R)\cos(t) - h\cos\left(\frac{r+R}{r}t\right),$$
$$y = (r + R)\sin(t) - h\sin\left(\frac{r+R}{r}t\right).$$

A model of this is shown in figure 10.13. The shape of this path is known as an epitrochoid, and to describe this further would take us too far from the topic of roundness to which this chapter is devoted. Its shape is described in Nash (1977) and for other related applications we highly recommend Holmes (1978).

10.5   Making Shapes of Constant Width

All curves of constant width are fascinating and can be cut from coloured plastic sheet or plywood easily enough. These can be made into rollers, or even used for decoration such as a mobile for a baby's room. Making a cylinder of wood or metal with the cross-section of a Reuleaux rotor is a very different story. To make *one* of these cylinders requires careful and precise marking out of the stock material and equally precise positioning of the work in a lathe. It is not impossible but we felt it to be well beyond our capabilities and so we have not even tried to make one. Instead we make *three* simultaneously using a very simple method.

One of the readily available stock sections of bar metal is a true hexagon. Aluminium is a good choice, as is brass. While they are easy to turn with standard equipment, the method is not obvious and we feel that a little more detail is in order. The first task is to go over all faces with a perfectly smooth flat file to remove any tarnish and to highlight the edges, which must not be touched. Cut off three lengths and hold these in a three-jaw self-centring chuck with no more than 60–70 mm protruding to be turned. Turn down with the lathe until the cuts just reach the edges, and do it slowly with fine cuts because the cut is interrupted.

Remove the work from the chuck and check that the jaws have not bruised the metal, restoring the flatness if necessary. Next, rotate each piece through 120° and repeat the process, and then repeat a third time before parting off. The ends of each cylinder can then be faced off. You now have three identical rollers and you have been able to achieve this without any measurement. The results of this procedure are shown in plate 22.

If you are fortunate enough to have access to a CNC lathe, turning solids that have a cross-section of any piecewise circular curve of constant width with an axis of symmetry is straightforward. Once the tool shape is known, programming the lathe for circular arc profiles is usually quite simple. This is how the aluminium solids in plate 18 were produced. There will always be a pip at the end of the solid, formed when parting off, but it can be filed off without difficulty and without spoiling the shape. Larger wooden solids can be made on a conventional screw cutting lathe by careful measurement.

When you have made a solid and satisfied yourself that it is of constant width, be prepared for some scepticism and disbelief when you show it to others. The usual demonstration is to place the solid on a table and roll a hardback book on it. In some positions the centre of mass will be moving upwards and the very slight extra effort can be sensed and interpreted as poor workmanship on your part because the width is variable. In positions where the centre of mass is falling the false conclusion is still apparent to the sceptic. For this reason it is more convincing if the solid is made from a low-density material, like wood or aluminium, rather than from brass.

10.6 Roundness

We have seen that shapes of constant width, such as Reuleaux's rotor, do have legitimate applications. However, more problematic are shapes with lobes which can crop up when manufacturing round parts. The production process may include imperfect machine tools that induce corresponding imperfections in the part being manufactured. In another situation, imagine holding a workpiece in a lathe using a three-jaw chuck. This work is compressed while it is turned, and when released the resulting shape may deform and hence deviate from roundness, even if it was manufactured to a high specification. If a workpiece is not sufficiently supported during turning, then it is possible to develop regular lobes. Another common engineering application in which lobes can develop is centreless grinding. In this process the workpiece is supported on a static blade between a high-speed grinding wheel and a slower-speed regulating wheel with a smaller diameter. Sometimes one or both wheels are aligned out of the horizontal plane. This imparts a horizontal velocity component to the work so that an additional feed mechanism is not necessary. There are advantages to centreless grinding, including shorter loading times. The workpiece is also being supported constantly, which makes it possible to either make heavier cuts or to grind shapes with a very small diameter. A normal lathe creates quite an axial load, which is absent in centreless grinding, and this is a crucial consideration for work that is easily distorted. The process allows automatic feeding and hence the continuous production of large quantities of small pieces. The simplicity of the machines cuts down maintenance costs. Despite these advantages, in centreless grinding a part is not held firmly and rotated on an axis. It is simply allowed to generate its own shape. Shapes of constant width can and do occur here. Hence, shapes with lobes do occur in manufacture and it is imperative to have practical tests that measure deviation from true roundness.

We often take improvements to manufacturing techniques for granted. It is only older readers who will remember seeing new cars with the instructions 'running in, please pass' in their rear windows. For the first 500 miles it was common practice to limit speed to 30 mph, and not to exceed 40 mph for the next 500 miles, before changing the oil. Improvements in the reproducibility and quality-control standards of the manufacture of rotating parts, in terms of roundness and surface finish, mean that a new car engine is effectively run-in as soon as it comes off the production line. It is as well to remember that car manufacture for most models is a mass production process.

Before we consider practical roundness tests in detail, we should mention two other aspects that are crucial to engineering practice. The first is *surface texture* and the second is *cylindricity*. Surface texture is concerned with the micro scale of the boundary of the shape, whereas we are most concerned here with the macro scale geometry. Of course, with very small departures from roundness and very large numbers of lobes on a surface it becomes impossible to form a clear distinction between the two. All manufactured objects are three dimensional, and hence one really makes a cylinder rather than a thin round disc. Since our interest in this chapter is the geometry of roundness of two-dimensional shapes, we shall concentrate exclusively on this and how we can measure departures from roundness. So, while the other two aspects cannot be ignored by a practical engineer we shall not comment further on them here.

When we describe a shape as 'round', we mean that the cross-section is circular, which is to say that all points are equidistant from the centre. Departure from roundness can be measured in a number of theoretical ways. The first is to find the maximum deviation from the minimum circumscribed circle. For example, if we carefully measure the width of a 20p piece, then the *width* is 0.844 in. However, a coin will not fit into a hole of this diameter. From our experiments, the minimum diameter hole into which a 20p piece will fit, i.e. the diameter of the minimum circumscribed circle, is 0.857 in. An experimental test block illustrating this is shown in figure 10.14. Despite appearances, the two holes have been machined to very high accuracy indeed, and an undamaged 20p piece only just fits into the bottom hole.

Figure 10.14. The minimum circumscribed circle of a 20p piece.

Similarly, we could measure the maximum deviation from the maximum inscribed circle. The *least squares reference circle* occurs where the area inside the shape but not the circle equals the area inside the circle but not the shape. Departure from roundness is the maximum deviation from this circle. Another measure that we will consider below is the *minimum zone circle*, which is defined as the minimum radial separation between two concentric circles positioned to enclose the shape. Note that only in special cases will the circumscribed and inscribed circles mentioned be those which give the minimum zone circle. All these are theoretical. An engineer needs some practical tests that can be applied in the workshop, from which these measures can be calculated.

10.7 The British Standard Summit Tests of BS3730

We have categorically seen that a shape that has constant width is not guaranteed to be a circle. What, then, can we measure to

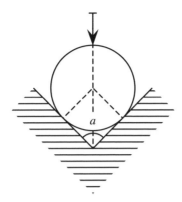

Figure 10.15. A circle in a vee-block.

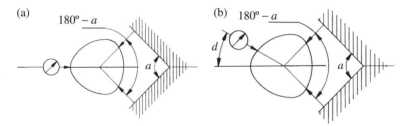

Figure 10.16. The British Standard summit methods:
(a) symmetrical and (b) asymmetrical.

ascertain if a shape is circular, and how can we quantify any departure from roundness?

We want you to imagine a circle resting in a vee-block with an internal angle a, as shown in figure 10.15. By symmetry the top of the circle will be found on the angle bisector of the two faces of the vee-block. We shall constrain a measuring device so that it remains on this line. It is clear that if the shape is rotated in the vee-block then the position of the measuring device stays constant and, furthermore, that this is true no matter how big or small the angle is. Can we conclude the converse? That is to say, if a shape rotates smoothly inside a vee-block and the position of the measurer remains constant, is the shape round?

Measuring the width in various orientations can be thought of as a *two-point measurement*. In engineering practice the two-point measurement is supplemented by various three-point methods. Indeed, part 3 of the British Standard for 'Assessment of departures from roundness' (BS3730, 1987, British Standards Institute) is concerned with 'Methods for determining departures from roundness using two- and three-point measurement'. BS3730 includes the so-called three-point summit method in symmetrical and asymmetrical settings: this is illustrated in figure 10.16. We will focus our attention on these tests. Other tests in the standard include the rider method, in which the workpiece rests in a vee-block and a measurement is taken between the vee-block and the workpiece. We do not consider these or methods for the measurement of internal surfaces, which are analogous.

The crucial question for us is the choice of angles a and d in the summit tests illustrated in figure 10.16. Quoting from BS3730 extensively:

> In order to cover all possible form deviations and numbers of undulations always take one two-point measurement and two three-point measurements at different angles between fixed anvils. NOTE: The measurement procedures may, under certain preconditions, be amplified (see Table 2, Table 3 and Table 4). Select the angles between fixed anvils from the following:
>
> - symmetrical setting: $a = 90°$ and $120°$ or $a = 72°$ and $108°$;
> - asymmetrical setting: $a = 120°$, $d = 60°$, or $a = 60°$, $d = 30°$;
>
> where a is the angle between fixed anvils; d is the angle between the direction of measurement and bisector of angle between fixed anvils.

We shall consider six tests, the first four of which are the symmetrical summit tests. We supplement these with a symmetrical test using a vee-block with an angle of 60°, which, strictly speaking, is not in the British Standard, and then consider one of the asymmetrical tests, which results in a closed triangle. In this triangle we have internal angles 30°, 60° and 90°. Table 10.1 shows the angles for the symmetrical tests together with the

Table 10.1. Angles in the BS3730 summit tests.

Test	a	$\dfrac{a}{360°} = \dfrac{p}{q}$	a'	$\dfrac{a'}{360°} = \dfrac{p}{q}$
A	72°	$\frac{1}{5}$	54°	$\frac{3}{20}$
B	90°	$\frac{1}{4}$	45°	$\frac{1}{8}$
C	108°	$\frac{3}{10}$	36°	$\frac{1}{10}$
D	120°	$\frac{1}{3}$	30°	$\frac{1}{12}$

angle $a' = \frac{1}{2}(180° - a)$. This is the base angle in an isosceles triangle formed from the angle a. We have also listed these angles as fractions of 360°, which will prove to be important later.

Some of the relevant data from the numbered tables in the standard that are quoted above are amalgamated in table 10.2. This table provides a correction factor, which we refer to as F, used in quantifying a departure from roundness for each of the tests A, B, C and D. Note that where the table shows a '—' the use of the specific test is prohibited when this number of undulations is known. For example, the column labelled 2 refers to the two-point measurement, i.e. measuring the width. The data show that a width test should not be applied to an object with an odd number of undulations, which in the light of our previous discussion is a necessary precaution. To obtain a measured departure from roundness, the formula

$$\delta = \Delta / F$$

is used, where Δ is the largest measured departure from roundness for the tests applied. Of course, this all assumes that the number of undulations is known in advance.

We should note at the outset that this portion of the standard does not claim to provide methods which always work accurately.

> The methods described in this Part of this standard may give faster and cheaper ways of assessing departures from roundness. This assessed value will deviate from the true value.
>
> BS3730, p. 4

Table 10.2. Correction factors for the British Standard summit tests (see text for details).

Number of undulations	2	72° A	90° B	108° C	120° D
2	2	0.47	1	1.38	1.58
3	—	2.62	2	1.38	1
4	2	0.38	0.41	—	0.42
5	—	1	2	2.24	2
6	2	2.38	1	—	—
7	—	0.62	—	1.38	2
8	2	1.53	2.41	1.38	0.42
9	—	2	—	—	1
10	2	0.70	1	2.24	1.58
11	—	2	2	—	—
12	2	1.53	0.41	0.38	2
13	—	0.62	2	1.38	—
14	2	2.38	1	—	1.58
15	—	1	—	2.24	1
16	2	0.38	2.41	—	0.42
17	—	2.62	—	1.38	2
18	2	0.47	1	1.38	—
19	—	—	2	—	2
20	2	2.70	0.41	2.24	0.42
21	—	—	2	—	1
22	2	0.47	1	1.38	1.58

Instead of perfect mathematical accuracy, the aim is more pragmatic since the vee-block test requires only relatively simple workshop equipment.

10.8   Three-Point Tests

We have conducted practical experiments with a standard 90° vee-block. For example, in figure 10.18 we have a 50p being supported by a vee-block. The vee-block and the dial gauge are

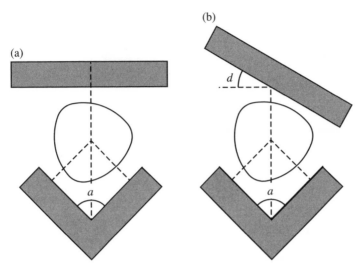

Figure 10.17. Three-point 'summit' methods:
(a) symmetrical and (b) asymmetrical.

supported on a very accurate plane surface and the position of
the dial gauge is moved to establish the maximum height of
the workpiece above the plane surface. Note that this is quite
a different procedure from fixing the vee-block and moving the
dial gauge vertically along the angle bisector between the two
fixed anvils, as would be the case in the tests specified in the
British Standard.

Our dial gauge measured vertical displacement to an accuracy
of ±0.01 mm. To this level of accuracy we could detect no difference between the point-up and point-down positions, or in
any of a number of positions in between. Indeed, the deforma-
tions caused by everyday use, such as being dropped on one corner, were larger than the difference caused by position. Hence,
according to the 90° vee-block alone a 50p was indistinguishable
from a round object. We shall see in a moment that the geometry suggests that we might actually expect this to be the case.
Indeed a 50p has seven undulations and looking at column B of
table 10.2 tells us that the 90° vee-block should not be used in
the British Standard tests either.

The Reuleaux rotor shown in plate 22 was quite a different
story. This was made to have a width of approximately 25.4 mm

Figure 10.18. Applying the three-point summit
method to a 50p piece with a 90° vee-block.

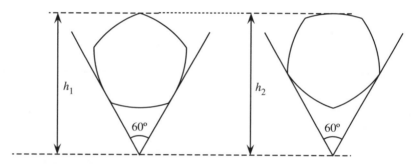

Figure 10.19. A circular arc rotor in a 60° vee-block.

(which is no accident), and the difference between the point-up
and point-down positions is 7.98 ± 0.01 mm.

We shall now use some trigonometry and apply this test to one
of the circular arc rotors to illustrate the technique. This time
we shall imagine a circular arc rotor generated from a regular
pentagon sitting in a 60° vee-block. Two potential positions for
this are shown in figure 10.19.

It is a straightforward exercise in trigonometry to show that if
the width is 1, then $h_1 = 1.497\,246$ and $h_2 = 1.502\,754$, giving
a difference of some 0.005 51 units, or 0.551%. Hence, this test
indicates that the shape is not round. But do such tests always
work? It certainly did not seem to work with a 50p in a 90° vee-
block. That is to say, our experiment did not establish that a 50p
is not round!

10.9 Shapes via an Envelope of Lines

To illustrate how the values in table 10.2 might arise we shall consider a problem in pure, but closely related, geometry. This is somewhat more technical than much of the rest of the book. Imagine that the three-point measurements now take place between straight anvils, two fixed and one moving. In particular, the two fixed anvils have an angle a between them as before. The workpiece to be measured rests on these anvils, and a measurement of the position of the third moving anvil is made. This is illustrated in figure 10.17. In the symmetrical test shown in part (a), an anvil is lowered onto the workpiece and as contact is made the relative positions of the three anvils are measured. In the asymmetrical test (part (b)), the third anvil moves along a line at an angle.

It is important to note a subtle, but crucial, difference between the tests proposed in the British Standard and those that we are about to discuss. We consider the distances between three *tangent lines*. In the case of a non-smooth shape, such as Reuleaux's rotor, these are support lines at fixed relative angles. In the British Standard summit tests there are two fixed anvils which, as before, operate as tangent lines. In the case of the symmetrical tests the measurement is taken by finding the *points on the curve* that lie on the angle bisector of these tangent lines. The point closest to the angle is the 'rider point', although we are most interested in that furthest away, which is the 'summit point' (see figure 10.16). We know nothing in particular about the direction of the tangent line at this point.

A shape of constant width can also be thought of as a rotor of the square. Indeed, this very property was used in both cam design and the square-hole drill, as demonstrated in plate 21. We are also interested in finding shapes of constant width that are also rotors of triangles. That is to say that the shape may be smoothly turned while remaining in contact with all three sides of a fixed triangle.

If, for example, we chose the angle a in the symmetrical vee-block to be 60°, we are led to consider the question, does there exist a shape with constant width which passes the 60° vee-block

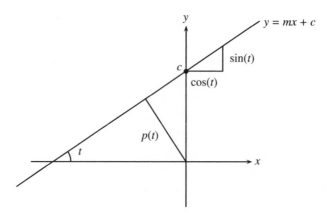

Figure 10.20. The equation of a straight line.

test? Such a shape would be a rotor of the square (constant width) and equilateral triangle (passes the 60° vee-block test). In fact we go further and explain how to construct a shape of constant width that rotates inside each member of a family of specified triangles. It will be clear that this shape is by no means unique, and that a whole family of such shapes exists.

Let us begin with the usual equation for a line, $y = mx + c$. We shall assume that this line makes an angle t with the x-axis and that the perpendicular distance of the line from the origin is $p = p(t)$. We prefer to make p depend on t, since in a moment we shall make extensive use of this. A diagram has been sketched in figure 10.20. Since m is the gradient of the line it follows that $\tan(t) = m$, and simple trigonometry allows us to see that $c = p(t)/\cos(t)$. Hence we can rewrite the equation of the line as $y = \tan(t)x + (p(t)/\cos(t))$, and this can in turn be recast in the form

$$y\cos(t) - x\sin(t) = p(t). \tag{10.1}$$

This form of a straight line is more general since vertical lines can be represented when $\cos(t) = 0$.

By giving a function for $p(t)$ we shall generate a family of lines, and we denote a particular line by L_t. The region enclosed is known as the *envelope* and this will enable us to generate a variety of shapes. When we talk of a *curve* we refer to the boundary of this region. This method of constructing curves as the envelope of a series of tangents has already been used in figure 7.10 to define the parabola using the purely mechanical means of a pin, straight edge and square.

Indeed, if we revisit figure 7.10, it is clear that the equation of the tangent line AP is

$$y = x \tan(t) - a \tan(t),$$

so that in the form (10.1) we have

$$y \cos(t) - x \sin(t) = -a \sin(t).$$

Hence, a parabola is generated by taking $p(t) = -a \sin(t)$. Another example of such an envelope of lines, showing a closed curve, is shown in figure 10.21.

For the purposes of this discussion we shall assume that t is measured in radians, so that an angle of 180° can more simply be written as π. Notice that by definition each line is tangent to the curve. Hence, to show that the curve has constant width it will be sufficient to consider the perpendicular distance between L_t and $L_{t+\pi}$, which are lines 180° apart, or rather parallels. In our case, the perpendicular distance between the two lines through the origin is simply $p(t) + p(t + \pi)$.

Here we will consider only envelopes generated by

$$p(t) = \alpha + \beta \cos(nt), \quad n \text{ is an odd integer} > 1. \qquad (10.2)$$

Notice that if $\beta = 0$ then all lines are a constant distance from the origin, and the curve generated will be a circle of radius α. We do need to be careful, however, since if we take β to be too large, then the resulting shape is not a convex region of the plane: see, for example, figure 10.22.

Here the lines L_t and $L_{t+\pi}$ will be parallel and a distance

$$2\alpha + \beta \cos(nt) + \beta \cos(nt + n\pi)$$

apart. Since n is an odd integer, $\cos(nt + n\pi) = -\cos(nt)$, so the lines are simply a distance 2α apart. Therefore, (10.2) automatically generates a shape of constant width.

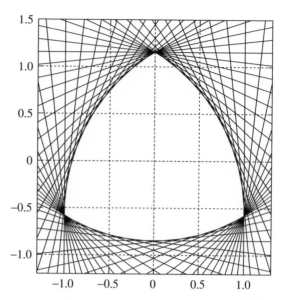

Figure 10.21. Reuleaux's rotor via an envelope of lines.

If we think of (10.2) as adding an odd-frequency perturbation to the circle, we might generalize this and add more than one such perturbation. Hence we could generalize (10.2) to something like

$$p(t) = \alpha + \beta_1 \cos(3t) + \beta_2 \cos(5t) + \cdots. \qquad (10.3)$$

It is even possible using an infinite sum to obtain an expression for Reuleaux's rotor, which Kearsley (1952) gives as

$$p(t) = \alpha + \sum_{n=1}^{\infty} \beta_n \cos(3nt),$$

where

$$\frac{\pi \beta_n}{3} = \frac{12n\alpha}{9n^2 - 1} \sin\left(\frac{\pi n}{2}\right) + \frac{4\alpha}{3n} \cos(3\pi n) \sin\left(\frac{\pi n}{2}\right).$$

A sketch of this is shown in figure 10.21. However, for us the generality of (10.3) adds nothing essentially new, and so here we consider only $p(t)$ given by (10.2). Of course, we still need to establish that the resulting envelope is a convex region of the plane.

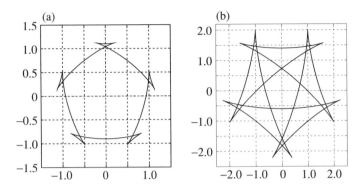

Figure 10.22. An example: $n = 5$ and $\alpha = 1$, with (a) $\beta = \frac{1}{10}$ and (b) $\beta = \frac{2}{5}$.

An envelope of lines is a very simple way of defining a shape, and it is also invaluable to know the equations of these tangent lines when cutting such a shape with a CNC milling machine. However, the collection of lines does not provide us with an explicit mathematical expression for the shape itself.

To derive a direct parametric representation of the curve we consider the point of intersection of L_t and $L_{t+\varepsilon}$. For epsilon very close to zero the lines are practically coincident, and the point of intersection practically lies on the curve itself. Taking the limit as ε tends to zero will be a point on the curve itself. A calculation gives this, and substituting this back for the y-coordinate gives an explicit parametric form for the curve:

$$\left.\begin{aligned}
x = {}& -\alpha \sin(t) + \tfrac{1}{2}\beta(n+1)\sin((n-1)t) \\
& + \tfrac{1}{2}\beta(n-1)\sin((n+1)t), \\
y = {}& \alpha \cos(t) + \tfrac{1}{2}\beta(n+1)\cos((n-1)t) \\
& - \tfrac{1}{2}\beta(n-1)\cos((n+1)t).
\end{aligned}\right\} \quad (10.4)$$

In this form it is clear that we have a circle of radius α to which two higher-frequency modes, with amplitudes proportional to β, have been added.

The parametric expression (10.4) allows us to derive straightforward criteria for when the envelope of lines generates a convex curve. This will occur if, and only if, there are no points of self-intersection of the curve. This is certainly satisfied if both $x(t)$ and $y(t)$ have at most two extremal values on the interval $t \in [0, 2\pi)$. Again, a calculation gives the criterion that

$$\alpha \geqslant |\beta(n^2 - 1)|. \tag{10.5}$$

Typically we shall assume that $\alpha = 1$ and take $\beta = 1/(n^2 - 1)$.

If we define $r^2 = x^2 + y^2$, then from (10.4) a further calculation gives

$$r^2 = \alpha^2 + \tfrac{1}{2}\beta^2(n^2 + 1) + 2\alpha\beta \cos(nt) - \tfrac{1}{2}\beta^2(n^2 - 1)\cos(2nt).$$

Taking $t = 0$ gives $r^2 = (\alpha + \beta)^2$, and taking $t = \pi/n$ gives $r^2 = (\alpha - \beta)^2$—these turn out to be the maximum and minimum values for r^2. Therefore,

$$\alpha - \beta \leqslant r \leqslant \alpha + \beta. \tag{10.6}$$

Measuring the departure from roundness using the minimum zone circle, we have a value of 2β.

So, we have not only a method for cutting this family of shapes on a CNC machine but also an explicit mathematical expression for the curve given by (10.4) and we know the departure from roundness.

10.10   Rotors of Triangles with Rational Angles

So far we have generated a family of convex curves. Now we try to see how we can choose the parameters to create rotors of triangles. In this section we begin by considering an arbitrary triangle, such as ABC shown in figure 10.23. We denote the lengths of sides BC, AC and AB by l_a, l_b and l_c, respectively, and the perpendicular distances from the origin p_a, p_b and p_c. By the sine rule we have that, for some k,

$$\frac{\sin(a)}{l_a} = \frac{\sin(b)}{l_b} = \frac{\sin(c)}{l_c} = k.$$

Then $\sin(a) = kl_a$, $\sin(b) = kl_b$ and $\sin(c) = kl_c$, and

$$2k \, \text{area}(\text{OBC}) = p_a \sin(a).$$

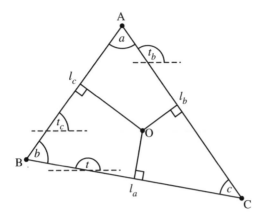

Figure 10.23. An arbitrary triangle.

Summing over the three triangles OBC, OAC and OAB gives

$$2k \, \mathrm{area}(ABC) = p_a \sin(a) + p_b \sin(b) + p_c \sin(c).$$

In addition, $2 \, \mathrm{area}(ABC) = l_b l_c \sin(a)$. Hence

$$2k \, \mathrm{area}(ABC) = l_c \sin(a) \sin(b).$$

Combining these results we have that

$$p_a \sin(a) + p_b \sin(b) + p_c \sin(c) = l_c \sin(a) \sin(b).$$

Take two similar triangles ABC and A'B'C' in which $l_a = \lambda l'_a$, etc. If

$$
\begin{aligned}
p_a \sin(a) + p_b \sin(b) &+ p_c \sin(c) \\
&= p'_a \sin(a) + p'_b \sin(b) + p'_c \sin(c),
\end{aligned}
$$

then

$$l_c \sin(a) \sin(b) = l'_c \sin(a) \sin(b) = \lambda l_c \sin(a) \sin(b),$$

so $\lambda = 1$ and hence ABC and A'B'C' are identical.

With this result in mind we make the assumption that $n \neq 1$ is an odd integer and define $\bar{n} = n \pm 1$, which is a non-zero even integer. We assume that angles a, b and c are such that $\bar{n}a$, $\bar{n}b$ and $\bar{n}c$ are all integer multiples of 2π. If we take a family of lines (10.2), then provided α and β satisfy (10.5) the curve is convex. We now argue that the curve is indeed a rotor of the triangle ABC.

We construct the triangle ABC by taking the lines with the angles $t_a = t$, $t_b = t - \pi + c$ and $t_c = t + \pi - b$. The triangles parametrized by t in this way will all be similar. We are required to show that they are identical, and hence the curve is indeed a rotor of the triangle ABC. To do this we note that $a = \pi - b - c$ and consider

$$p(t)\sin(a) + p(t_b)\sin(b) + p(t_c)\sin(c)$$
$$= \alpha(\sin(a) + \sin(b) + \sin(c))$$
$$+ \beta\sin(\pi - b - c)\cos((\bar{n} - 1)t$$
$$+ \beta\sin(b)\cos((\bar{n} - 1)(t - \pi + c))$$
$$+ \beta\sin(c)\cos((\bar{n} - 1)(t + \pi - b)).$$

Omitting a lengthy calculation in which we use the assumptions that $\bar{n}$ is a non-zero even integer and that $\bar{n}b$ and $\bar{n}c$ are multiples of 2π we arrive at

$$p(t)\sin(a) + p(t_b)\sin(b) + p(t_c)\sin(c)$$
$$= \alpha(\sin(a) + \sin(b) + \sin(c)),$$

which is independent of t, and hence constant.

10.11 Examples of Rotors of Triangles

First we apply these results with the modest goal of designing a shape of constant width that is also a rotor of the equilateral triangle. All the angles in an equilateral triangle are 60°, or $\pi/3$. Hence, taking $\bar{n} = 6$ ensures that $\bar{n}$ multiplied by each angle is equal to 2π. Hence $n = 5$ is sufficient in (10.2), although alternatively we could have taken $n = 7$. If we take $\alpha = 1$ then (10.5) gives the largest possible $\beta = \frac{1}{24}$, and each line in the envelope will be of the form

$$y\cos(t) - x\sin(t) = 1 + \tfrac{1}{24}\cos(5t). \tag{10.7}$$

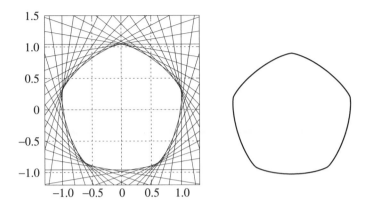

Figure 10.24. A rotor of both the square and the equilateral triangle.

An envelope of forty lines is shown on the left of figure 10.24, and its parametric expression

$$x = -\sin(t) + \tfrac{1}{8}\sin(4t) + \tfrac{1}{12}\sin(6t),$$
$$y = \cos(t) + \tfrac{1}{8}\cos(4t) - \tfrac{1}{12}\cos(6t)$$

$$(10.8)$$

is plotted on the right. In this case, the width is 2 and a calculation shows that the lengths of the sides in the triangle are $2\sqrt{3}$. Taking concentric circles with radii $\alpha \pm \beta$ gives the departure from roundness (using the minimum zone circle measurement) of $\tfrac{1}{12}$, or approximately 8.3%.

Next take a more general triangle in which all the angles are rational fractions of a rotation. That is to say, each angle a can be written as

$$\frac{a}{2\pi} = \frac{a}{360°} = \frac{p}{q}.$$

Assume that for each angle in the triangle $\bar{n}(p/q)$ is an integer. Let n be defined to be either $\bar{n} + 1$ or $\bar{n} - 1$. Assume further that the convexity constraint (10.5) is satisfied. Then the scheme (10.2) will generate a rotor of this triangle. Taking $\bar{n}$ as the lowest common multiple of the denominators q suffices.

If we consider an isosceles right-angled triangle, this has two different angles 90° and 45° which can be expressed as a $\tfrac{1}{4}$ and an $\tfrac{1}{8}$ of a rotation. It is clear that taking $\bar{n} = 8$ ensures that

$\bar{n}(p/q)$ is an integer, and so $n = 7$ or $n = 9$ generates a rotor of the isosceles right-angled triangle.

If we want to create a shape that will be a rotor of more than one triangle, then we need to consider all the angles involved. For example, to find a shape that is simultaneously a rotor of the equilateral triangle and of the isosceles right-angled triangle, we now have three fractions to consider: $60° = \frac{1}{6}$, $\frac{1}{4}$ and $\frac{1}{8}$. Taking $\bar{n} = 24$ and $n = 23$ suffices.

We might go even further and try to find a single shape that is a rotor of all the triangles with angles specified in table 10.1. We take $\bar{n}$ as the lowest common multiple of the denominators of fractions a' in this table. Hence $\bar{n} = 8 \times 3 \times 5 = 120$ suffices for the triangles generated by the angles in table 10.1. If we supplement these tests with a $60°$ vee-block, resulting in an equilateral triangle, then $a = a' = 60°$ and $a/360° = \frac{1}{6}$. We refer to this as test E and in fact $\bar{n} = 120$ also generates a rotor of the equilateral triangle. More importantly, for the asymmetrical case $a = 60°$ and $d = 30°$ we generate a triangle with internal angles $60°$, $90°$ and $30°$, all of which are accounted for already. Hence, this asymmetrical summit test adds nothing essentially new to this geometric problem. The remaining asymmetrical test does not result in a triangle, and we do not consider it further.

Taking $n = \bar{n} - 1 = 119$ in (10.5) suggests that this generates a departure from roundness of about $2/(119^2 - 1) = \frac{1}{7080}$, or 0.014%. This is too small for us to be able to provide an illustration, but it would be similar to that in figure 10.24 with 119 undulations.

If we choose the pairs of triangles that result from legitimate choices in the British Standard, then the first option for symmetrical settings is to apply tests B and D. Doing this results in fractional angles without a factor of 5 in the denominator. Hence $\bar{n} = 24$ suffices and consequently the departure from roundness is 0.38%. The second option for the symmetrical settings is to apply tests A and C. Any shape generated by (10.1) which is a rotor of the triangle with angles specified by test A is automatically a rotor of the triangle with angles from test C, and so from our point of view this test contributes nothing and $\bar{n} = 20$ suffices, giving the departure from roundness as 0.56%.

If we return to our experiment with the 50p piece then we have a shape with 7 undulations—again it makes little sense to apply test B, as we did in figure 10.18. With $\bar{n} = 8$, the test ignores $n = 7$ undulations. Interestingly, if we consider test A from table 10.1, then $\bar{n} = 20$ is the smallest value and our related geometric analysis would suggest that we should beware of using this test with $n = \bar{n} \pm 1$, i.e. 19 or 21 undulations. Indeed, these are precisely the numbers of undulations for which the British Standard summit test A should not be used. For test B, $\bar{n} = 8$. Here we need to consider $n = k\bar{n} \pm 1$ for $k = 1, 2, \ldots$, since after all we only require $\bar{n}(p/q)$ to be an integer for each angle in the test. Thus we should avoid this test for shapes with 7, 9, 15 or 17 undulations. Although our tests are not those specified in the British Standard, the geometry we have examined goes some way to explaining how table 10.2 might be derived.

One might imagine that by applying the two-point measurement method and then a number of different three-point techniques we could be confident that the shape was round. Indeed, Reason (1966, p. 17) claims the following.

> If several V-block determinations are recorded each with a different angle, it should be possible, with an electronic computer, to find the true shape of the part from the combined results, the principle being that there can only be one set of radial measurements that will account simultaneously for all the V-block measurements.

Our method allows us to create a rotor for a whole family of triangles. Hence, it seems that Reason's principle provides very shaky foundations indeed, especially when a finite number of vee-blocks are used. It appears that from a purely theoretical point of view our vee-block tests are little better than width measurement.

No matter how many different tests we propose applying, provided all the angles are rational, we can construct a single shape which is a rotor of them all. However, our test involving a triangle with some *irrational* angles cannot be accommodated within this scheme. Indeed, a remarkable theorem of Kamenetskiĭ (1947) applied to triangles proves that necessary and sufficient conditions for the existence of a rotor of a triangle are that all angles are rational. Thus, a *single* irrational triangle has no rotors, and hence would provide a sufficient three-point summit test for roundness. While constructing an irrational-angled triangle might not be practical for the engineer, using fractional denominators that have a large lowest common multiple would result in a much greater $\bar{n}$ and hence less deviation from roundness.

It is precisely this approach that is rather cleverly adopted by Goho, Kimiyuki and Hayashi (1999). They propose an angle of $14° = \frac{7}{180}$. In a symmetrical summit test the other two angles in this triangle are $\frac{1}{2}(180° - 14°) = 83°$. Since 83 is prime, $83/360°$ cannot be simplified and we would need to take $\bar{n} = 360$, giving $n = 359$. With $\alpha = 1$, the convexity constraint of (10.5) gives

$$\beta = \frac{1}{n^2 - 1} = \frac{1}{129\,599}.$$

Hence, the departure from roundness is $\frac{2}{129\,599}$, or approximately 0.0015%, which is rather small. Furthermore, while we have chosen frequency modes in (10.2) in such a way that peaks and troughs cancel out so that there is no apparent difference in height of the dial gauge, other modes are actually *amplified* by rotation in a vee-block. This is shown in table 10.2, in which the F factors are greater that one. When the angle a is 14°, this can be by a factor of about nine. The practical engineer might well be happy knowing that some very-high-frequency modes, which result in small-amplitude departures from roundness, are ignored if the larger lower-frequency modes are amplified.

10.12   Modern and Accurate Roundness Methods

To complete this chapter we will very briefly describe how roundness is determined. We have comprehensively shown that simple vee-block methods do not provide reliable tests. One intuitive reason for this is that three points of contact contribute to the measured value in each position. This provides the opportunity for lobes to generate peaks and troughs that cancel each other out. Indeed, by considering regular lobes it is possible to derive an intuitive notion of when the three-point measurement is likely to give misleading results. While an irrational-angled triangle might be made to work, a much more direct method is to establish the position of the edge of the shape with a single point of contact.

We should recall that a round shape has a circular cross-section, and hence all points on the edge are equidistant from an axis of rotation. To determine roundness accurately a machine incorporates a turntable to rotate the object about a reference axis. This axis is independent of the part being measured. Some kind of measuring device, in the form of a stylus, then presses against the surface as the object rotates and a transducer converts the motion of the stylus into positional information. Typically a system will use 3600 points or more.

However, there are real practical difficulties in setting up the machine prior to such a measurement being taken. The first is the problem of centring. Consider, for example, measuring a sphere rotated about an axis that does not pass through the centre of the sphere. The next problem is levelling. To illustrate this consider measuring a perfect cylinder that has been tilted. A plane section is basically elliptical. Furthermore, if the stylus tip has a finite size, then the contact point also moves up and down the cylinder as it is rotated. This further distorts the measured shape—a problem known as cresting. Here, the projected line of action of the gauge does not pass through the axis of rotation. While cresting can be eliminated by correctly setting up the stylus, centring and levelling need to be performed for each measurement.

Modern machines can automatically move the object into position and establish the position and orientation of the object's axis. They will then determine the departure from roundness using this axis. Such machines incorporate a large number of moving parts, all of which must be machined to the very highest specifications if imperfections in their manufacture are not to induce corresponding imperfections in the departure-from-roundness measurements being taken. These moving parts, the software needed to control such machines, and the calibration all combine. Highly expensive specialist equipment is the result. This complexity is necessitated by the geometry we have explained in this chapter.

Chapter 11

PLENTY OF SLIDE RULE

Yet I confess I like it the better, because it pleaseth not
your palate, to which nothing can savour, that is learned and
Analyticall: but onely the superficiall scumme and froth of
Intrumentall trickes and practices.

Oughtred (1634)

We want to ask you to imagine you are the captain of a super-
tanker that is docked in port and being loading with oil. Your
tanker has a number of separate tanks, perhaps of different
volumes, and oil is being pumped aboard. If you decide to fill
one tank at a time then the ship will capsize, so you need to
balance the ship by part-filling tanks around the vessel. Oil varies
in density, and so you cannot rely on a set of pre-prepared tables
since you may not know how heavy the oil is until you start
the job. What you need to help you do this is a simple, reliable
and quick method of performing these volume, density, flow rate
calculations on the deck of your ship.

There has always been a need to perform such calculations:
they were used by ships' captains, carpenters and tax collectors,
among others. Remember that space flight predates by some
decades the cheap mass-produced electronic calculator. Most
modern computers are digital; however, there are a number of
devices for calculating that rely on continuously changing physical quantities. Such devices are known as *analogue computers.*
These quantities might be water flow rates, electrical voltages or
the physical position of a mechanism.

Such devices were widely used for practical calculations which
did not need more than two or three significant figures of accuracy, and it is precisely one of these that William Oughtred

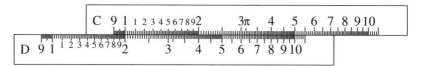

Figure 11.1. Setting two logarithmic rulers to perform 2×3.

dismisses above as 'the superficiall scumme and froth of Intrumentall trickes'. Indeed, the quotation is one salvo in a bitter
argument over the priority for the invention of analogue devices
that use *length* and *angle* as the physical quantities. These are
known as slide rules, and they rely on the relative physical positions of two scales. Such devices are the subject of this chapter.

Have a look at the rules shown in figure 11.1, which are set
up to perform the multiplication 2×3. The first thing to notice
is that the scales themselves are not equally spaced. Actually
they are drawn using *logarithms* and reading a logarithmic scale
does take a bit of practice. Notice that the distances between the
marks are uneven and that there are more marks between 1 and
2 than between 9 and 10. If you count, you will find that between
1 and 2 there are forty-nine marks but between 9 and 10 there
are only nine.

We shall explain how to use the slide rule using examples,
starting with the simple product 2×3, which may be calculated
by using the two logarithmic rulers shown in figure 11.1. The
two scales have been labelled C and D as these letters are traditionally used on slide rules for the scales shown here. Place
them side by side so that the 1 on C is next to the 2 on D. Now
look on C for the 3. The answer to the product 2×3 is the value
that is opposite this 3 on the D scale, as shown in figure 11.1.
What we have actually done is added the logarithm of 2 (on the D
scale) to the logarithm of 3 (on the C scale), and this process will
be fully explained in section 11.1. Notice that *without moving
the ruler* we may calculate lots of other products. For example,
we can read 2×4 and 2×5 off the slide rule without touching
either ruler. This feature actually makes the slide rule quicker
to use than the electronic calculator for some repetitive tasks,
such as converting a student's homework mark to a percentage,
at a glance.

11.1   The Logarithmic Slide Rule

Just as in the above example, the most common use of a slide rule is to perform multiplication (or division) based on the laws of logarithms. If $x > 0$ then the logarithm of x to the base $a > 0$ is the unique number y that satisfies

$$a^y = x.$$

This is written as $y := \log_a(x)$. For example, $10^3 = 1000$, so $\log_{10}(1000) = 3$, and $2^6 = 64$, so $\log_2(64) = 6$.

From this it follows that

$$\log_a(1) = 0 \quad \text{and} \quad \log_a(a) = 1 \quad \text{for all } a > 0. \qquad (11.1)$$

In practice a is taken to be 10 or the special number e $\approx$ 2.71828. This latter base gives what are called the natural logarithms, which become important mathematically from calculus onwards. They are often denoted by $\ln(x)$ rather than $\log_e(x)$.

Taking $a = 10$ is usual in practical calculation because of the following link with number bases in the place–value number system. If we work through the above definition of logarithm it is straightforward to see that

$$\log_{10}(10y) = 1 + \log_{10}(y).$$

Now, repeated use of this together with (11.1) gives

$$\log_{10}(100) = 2, \quad \log_{10}(1000) = 3, \quad \dots, \quad \log_{10}(10^y) = y,$$

and so $\log_{10}(\frac{1}{10}) = -1$. Often we simply write log in place of $\log_{10}$ where there is no risk of ambiguity.

The crucially important property of logarithms derives directly from the law for exponents that

$$a^y a^z = a^{y+z},$$

and they allow us to convert a multiplication into addition using the relationship

$$\log(y) + \log(z) = \log(y \times z) \quad \text{for all } y, z > 0. \qquad (11.2)$$

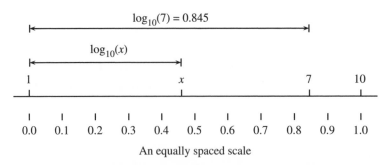

Figure 11.2. Drawing a logarithmic scale.

The invention of the logarithm by John Napier in 1614 was perhaps the most important advance in practical calculation since the introduction of the place–value (decimal) number system. Napier was a colourful and inventive character. Amongst his other writings, Napier published *A Plaine Discovery of the Whole Revelation of Saint John* in English in 1593. This contained proofs, in the style of Euclid, that the pope was the Antichrist and that the world was due to end in 1786. What we remember Napier best for is the small book, first published in 1614, entitled *Mirifici Logarithmorum Canonis Descriptio* ('Description of the Wonderful Canon of Logarithms'). This book was a set of tables and a short description of how to use them. An explanation of the theory and the method of calculation was published posthumously in 1619 as *Mirifici Logarithmorum Canonis Constructio* ('Construction of the Wonderful Canon of Logarithms'). This was translated subsequently by Henry Briggs and published in English. It should be noted that Napier's concept of logarithm was quite different, although of course mathematically related, to (11.2). More details of this may be found in, for example, Edwards (1979).

Although logarithms now appear modestly as a button on an electronic calculator, their effect on the seventeenth-century scientific community was profound. In Napier's time, the concept of a fractional exponent had not been developed and neither had the general concept of a function—it would therefore have been impossible for Napier to have given the synopsis of logarithms that we have done. Hence Napier's definition differed from ours and was based on the continuous movement of two points. Without calculus this was the only way of considering continuously varying quantities. When Johannes Kepler (1571–1630) received a copy of Napier's tables, he used them to assist him with the calculations that led to the formulation of his laws of planetary motion. Without the aid of logarithms these calculations would have taken many years, and Kepler published a letter addressed to Napier as the dedication to his *Ephemeris* of 1620 congratulating him. In turn, Kepler's laws gave Newton evidence that was crucial to support his theory of universal gravitation. With this in mind we get some idea of the significance of Napier's discoveries.

To implement the mathematical law (11.2) physically, we need to draw logarithmic scales. On such a logarithmic scale the physical length from the origin is proportional to the logarithm of the number marked on the scale. For example, $\log(1) = 0$ and so the 1 is marked at the origin of the scale. To make a logarithmic scale, first draw a line and mark two points on it. These will correspond to the numbers 1 and 10. To mark a point on the scale, 7 for example, we measure a length proportional to the logarithm of 7 along the line. This is illustrated in figure 11.2. A completed logarithmic scale is show in figure 11.3.

In chapter 4 we made much of the difficulty in accurately constructing a scale by pure geometry, without any *measuring* whatsoever. Unfortunately, apart from the ends 0 and 1, none of the positions of the other points can be established by construction. The procedure undertaken by practitioners such as Stone (1753) began with dividing a scale up into equal parts, perhaps 1000. Once this was complete a table of logarithms was used to find the position of each mark. The positions of prime numbers could be used to establish the other marks. For example, if we mark out the positions of 2 and 3, addition of the physical lengths by geometrical construction allows us to find the positions of 4, 6, 8, 9 and so on. However, this procedure risks the introduction of cumulative errors when the slide rule is being marked out by hand by a craftsman. Rules would originally have been marked out in this way but craft production eventually gave way to mass production.

Figure 11.3. A logarithmic scale.

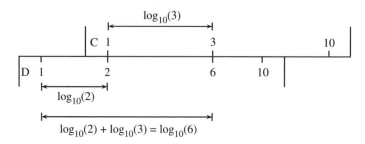

Figure 11.4. The principle of a slide rule.

In the opening section of this chapter we calculated the product 2×3 by using two logarithmic rules. In effect we had added the lengths, and hence added the logarithms of the numbers. Since (11.2) assures us that adding logarithms corresponds to multiplying the numbers, we can now see the principle of the logarithmic slide rule. This calculation is shown again in figure 11.4 to emphasize the addition of physical lengths.

Although it is possible to perform many other similar calculations immediately, it is impossible to calculate 2×6 because we run out of scale. If you ever run out of scale like this you need to *divide by ten*. This corresponds to moving the whole of the C ruler to the left so that the 10 on the C is where the 1 currently is, as shown in figure 11.5. The 6 on C is now above 1.2 on D but we *divided by ten* so the answer must be 12, which is, of course, 2×6. Similarly we can read off 2×7, 2×8, etc., the scale reading being multiplied by ten.

It is possible to design a circular slide rule in which the angle, not the length, is proportional to the logarithm—we shall examine circular rules in section 11.4. The problem of insufficient scale does not occur with a circular pattern rule, although the position of the decimal point must still be established by mental calculation. In fact, when using a slide rule the decimal point must always be established by mental arithmetic. Hence, the

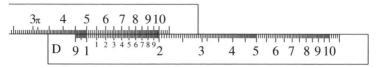

Figure 11.5. Setting two logarithmic rulers to perform 2×6.

rules in figure 11.1 may just as readily be used to calculate 2×3, as 200×0.03.

Division may be performed equally simply, essentially using the reverse process by noting that (11.2) may be rewritten in the form

$$\log(y) - \log(z) = \log\left(\frac{y}{z}\right) \quad \text{for all } y, z > 0.$$

If we begin this time with y on the D scale and locate z on the C scale, the difference between these two logarithms will be the logarithm of y/z. Figure 11.4 can be reinterpreted to show the rule set-up for 6 divided by 3. Look at figure 11.5: in this, the 7 on the C scale is above the 1.4 on the D, so this configuration is precisely that needed to calculate $14 \div 7$, and the 10 on C is indeed above the correct answer, i.e. 2, on the D scale.

The basic principle of the slide rule is simple—however, there are a number of subtleties in its practical operation that require skill together with manual and mental dexterity. Next we examine the history of the slide rule and then, in section 11.3, we explain how other common calculations are performed, often with the aid of different scales. This only scratches the surface of an extensive field, albeit a rapidly dying one.

11.2 The Invention of Slide Rules

The slide rule is an extremely good case study of how the development of technology progresses in small incremental steps rather than by large leaps. Originally, a single logarithmic scale was inscribed upon a rule by a craftsman using a table of logarithms to establish the positions of the individual marks. A pair of dividers was used to walk along this single scale to add lengths. Next, two rulers were placed side by side on a desk, just as in figure 11.1. From this a gradual refinement of the tool took place with the introduction of a sliding cursor, better scale layout and new materials. Every step represents a genuine improvement, although each one by itself seems rather unimportant. The sum total is a sophisticated, accurate and highly versatile instrument. One could expect a late-twentieth-century general-purpose slide rule to be inscribed with up to thirty different scales aligned on both sides.

The first scientific instrument to contain a logarithmic scale was the sector, which was discussed separately in section 4.6. This innovation is credited to Edmund Gunter, who became Gresham Professor of Astronomy in 1619. His most important book was his *Description and Use of the Sector*, which was first published in English in 1623. This was hailed by Cotter (1981) as 'the most important work on the science of navigation to be published in the seventeenth century'. Now, Gunter's sector is not a slide rule in any sense of the term. It contains a single logarithmic scale that is used in conjunction with a pair of dividers. Imagine that we wish to perform a multiplication calculation, for example. Begin by opening the dividers a distance corresponding to the logarithm of one number, as measured on the scale. Then place one point on the scale itself at the point corresponding to the other number and see where on the scale the other point comes into contact. We have added lengths by using the legs of the dividers.

The actual invention of the slide rule came next, and like many of the devices described in this book, its priority for invention is a complex issue, involving a controversy between two rival inventors: William Oughtred (1573–1660), whose detailed biography is Cajori (1916), and Richard Delamain (1610–45) (see Taylor 1954, section 122). It is generally accepted that Oughtred was the inventor of the first slide rule but that Delamain published first, and, furthermore, that the two probably independently invented their instruments (see Cajori 1920, p. 205). The first published account of a circular slide rule is to be found in Richard Delamain's 30-page pamphlet *Grammelogia* of 1631. This was closely followed by William Oughtred's *The Circles of Proportion and the Horizontall Instrument* (Oxford University

Press, 1632). Note that the 'horizontall instrument' is a form of sundial not slide rule (Turner 1981). Translated and published for Oughtred by William Forster (Taylor 1954, section 164), the introduction attacks Delamain. Delamain's expanded 113-page *Grammelogia* of 1632 replies to the attack contained in the introduction to *Circles* equally forcibly. In 1633 Oughtred published an edition of *Circles* with a section titled 'Hereunto is annexed the excellent Use of two Rulers for Calculation'. This details the linear slide rule for which Oughtred has undisputed priority. The 1634 edition of *Circles* contains the 32-page section with the following title.

> TO THE ENGLISH GENTRIE and all others studious of the MATHEMATIKS, which shall be *Readers* hereof. The just Apologie of WIL: OUGHTRED, against the slanderous insimulations of RICHARD DELAMAIN, in a Pamphlet called *Grammelogia*, or *Mathematicall Ring*, or *Mirifica logarithmorum projectio circularis*.

This section is referred to as Oughtred's *Apologie* and we shall examine the argument between Oughtred and Delamain in more detail in section 11.9, since their views on using calculating instruments in teaching are fascinating, and still relevant.

The further development of the slide rule cannot really be isolated from that of other scientific instruments. Indeed, all instruments of this period were hand inscribed, making them very expensive and possibly inaccurate. Also, instrument makers in the early seventeenth century understandably guarded their trade secrets very carefully. Such secrecy not only reduced communication of ideas but inevitably led to independent reinventions. This sparked controversy, although both Oughtred and Delamain used the same instrument maker, Elias Allen. It was not until 1753 when Edmund Stone published his *Construction and Principle uses of Mathematical Instruments* that such information began to become available. The work of Hopp (1998) divides the history of slide rule manufacture into approximately two periods. The first was from the 1630s to the 1850s. This is characterized by individual instrument makers, each with their own patterns for the scale layouts and innovations in the physical design. From the 1850s onwards the patterns of scales became more standardized and manufacture and mass production became more common. The introduction of better materials, specifically printed celluloid scales and plastics, brought the costs down and improved accuracy, making the slide rule a useful and affordable tool. We shall sketch out some of these important innovations below.

Although an apparently obvious idea, it seems that it was not until 1654 that Robert Bissaker thought of placing a slide between two fixed stocks rather than simply placing two rules side by side on a desk. This is the first recognizable slide rule, although the name does not appear until 1662 when John Brown used the term in a book. During the later seventeenth century standard designs began to emerge, such as those of Coggeshall and Everard.

Physical innovations continued, as did different arrangements of scales until the pattern most commonly found on modern slide rules was first invented by Amédée Mannheim (1831–1906) around 1850. His rule also contained a cursor, which allows one to read scales that are not contiguous accurately. Although this was invented in England two centuries earlier by Newton, on the device examined by Sangwin (2002), it only became popular with Mannheim's rule.

Mannheim's rule is based around two logarithmic scales, one on the slider and one on the bottom of the stock, which are traditionally labelled C and D, respectively. In addition, two other scales marked A and B contain double logarithmic scales. That is to say, if a reading x is taken on the D scale, directly above this is the reading x^2 on the A scale. The other most common scales are K, a triple scale at the top of the rule, and the scale CI on the slider, which is an inverse scale. Trigonometrical relationships are generally given on the reverse side of the slider.

Following Mannheim's rule innovations continued, including the addition of a magnifying glass to the cursor. Different physical designs became popular, including pocket watch circular calculators and cylindrical calculators. Other scales such as the

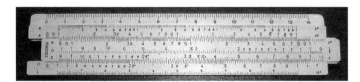

Figure 11.6. A typical slide rule, the Faber Castell 67/87 Rietz.

log–log scale were introduced. The duplex rule—that is, a rule with scales aligned on both sides—was patented in 1891 by William Cox (US Patent 460930). The folded scales, which will be explained below, were introduced by Auguste Beghin in 1898. Development and mass manufacture of slide rules continued until the late 1980s, with the last rules produced designed for special applications. Some rules, such as the *Moonstick*, which as the name implies is used for calculations involving phases of the Moon, are still available.

11.3   Other Calculations and Scales

In addition to two simple logarithmic scales, an early twentieth-century slide rule traditionally came equipped with a number of others. These are designed to facilitate a variety of calculations, including trigonometrical ones. In this section we look at just a couple of the more common scales to give you some idea of the range possible. Most slide rules came with instruction manuals, which have the advantage of relating example calculations to the actual scales on the rule. In addition, there are many general instruction manuals: see, for example, Dobson (1972) and Snodgrass (1959).

In the multiplication calculations illustrated in figure 11.1 we soon ran out of scale, i.e. 2×6 cannot be found. One solution to this problem was first used by Auguste Beghin in 1898 and is known in its modern form as the folded scales. To use the folded scales we need to move away from two juxtaposed scales and imagine a number of different scales on both the fixed stock and the moving slider. In fact we imagine that the C and D scales are continued to the right. This would evidently solve the problem. The extensions to the C and D scales are cut off, and labelled CF

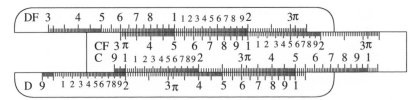

Figure 11.7. Using the scales CF and DF to perform 2×6.

and DF, respectively. These two new scales are displaced physically back to the left, so that the C and CF scales are on the slider and move together. The D scale is contiguous to C, usually on the bottom of the stock, whereas the DF scale is contiguous to CF on the top of the stock.

However, the CF and DF scales do not begin at 1. In fact, they begin about halfway along the physical scale. Mathematically, half the physical distance along the scale corresponds to $10^{1/2} = \sqrt{10} \approx 3.1623$. Actually, a more popular displacement for the folded scales is $\pi \approx 3.1416$, which by chance is approximately $\sqrt{10}$, but more useful in calculations.

Since the C and CF scales are fixed together on the slider and the D and DF scales are fixed together on the stock, any calculation performed on the C/D scales will also be performed on the CF/DF scales. However, the portion of the scales that is visible will be different. So, if we return to our calculation of 2×6, we cannot read the answer by looking below the 6 on the C scale because we have run out of D scale. Instead, we can look above the 6 on the CF scale and find the 12 on the DF scale. This is illustrated in figure 11.7. In ways similar to this, we can use combinations of the C/D scales and the CF/DF scales to ensure that the result appears on one of the scales. This is difficult to describe in words, but is actually evident with a little practice.

Two other scales that commonly appear on slide rules are labelled A and B, with A fixed on the stock and B on the slider. Both of these fit two complete scales into the physical distances occupied by the C and D scales. Therefore, by the laws of logarithms, the A scale shows the square of the D scale, and similarly the B scale the square of the C. Since squaring and taking the square root of a number are very common functions, the usefulness of this is evident. However, since the A and B scales occupy only half the physical space, they have only half the accuracy. In using these scales for squares and square roots, real care is needed in determining the position of the decimal point.

The real power of this arrangement is when calculations on the C/D scales are combined with calculations on the A/B scales. To ensure the maximum overall accuracy, it is important to minimize the total number of slider movements and intermediate calculations. To achieve this a calculation must be planned in advance. The cursor can be used to 'store' an intermediate result, while the slider is moved. This removes the need to note down such a result on paper. On a late-twentieth-century rule both sides are used and the scales on each side are aligned. On both sides the C/D scales appear, with A/B scales on one side and CF/DF scales on the other. By turning the rule over and using both sides simultaneously in a calculation, the best use can be made of all the scales. In doing this the cursor is invaluable for transferring a number from one side to the other.

Some numbers, such as π, occur regularly in calculations. As in figure 11.7, such values are marked on a rule for speed and accuracy and are known as *gauge points*. Gauge points are specific values that are marked on the rule to help with specific calculations. Two other common gauge points are marked ϱ' and ϱ'': these show, respectively, the factors for multiplying minutes and seconds of arc to turn them into radians. Which other gauge points are present depends very much on the intended user of the rule. For example, carpenters, tax officials and accountants all had their own gauge points for helping with calculations relevant to their professions.

Another common feature on slide rules is an extra hairline on the cursor of the rule. This is offset by a fixed distance and allows for a multiplication factor to be immediately calculated. The most common auxiliary hairline appears to the left of the cursor hairline for use with the A/B scales. This corresponds to a multiplication factor of $\pi/4 \approx 0.785$. Since the area of a circle is $(\pi/4)d^2$, this auxiliary hairline is invaluable for this common calculation. To perform this, move the cursor to the diameter on the D scale, and read the square of this on the A scale above. Then, looking at the position of the auxiliary hairline, find the area on the A scale. Note that for this calculation the slider is not needed. Similarly, a tax inspector might have an auxiliary hairline at 1.175 to allow addition of a tax of 17.5% very accurately and swiftly.

There are also scales for trigonometrical calculations. For example, if one needs to calculate $a \times \sin(t)$, then one adds the length corresponding to $\log(a)$ to that corresponding to $\log(\sin(t))$, which may be read directly from one of the other scales. Unlike an electronic calculator that copes with the periodicity of the circular functions, the user needs to do much more when using a slide rule. The sine of an angle is usually only given for $0 \leqslant t \leqslant 60°$, and a user would be expected to convert values of t into this range before picking up the rule. This can be done by using trigonometrical relationships such as $\sin(90° - t) = \cos(t)$.

11.4   Circular and Cylindrical Slide Rules

The linear slide rule is only one possible layout for the scales. Indeed, Oughtred's original design was circular. Although significantly less common than the linear slide rule, both circular and helical cylindrical slide rules were produced. The helical slide rule in particular has a much greater scale length and hence is far more accurate than its linear counterpart. A circular slide rule with conversion tables, the 'Concise Type E', is shown in figure 11.9.

The most significant advantage of the circular design is that one does not run out of scale. The calculation continues around the circle, with the user keeping track of the decimal point as with the linear rule. Circular designs have another advantage: by wrapping a scale round a circle we get just over three times as much scale on a circle of given diameter as we do on a straight rule of that length, and the helical slide rules had an even greater scale length. For example, the Otis–King calculator, which was patented in 1922, has a scale length of 66 in wrapped around a

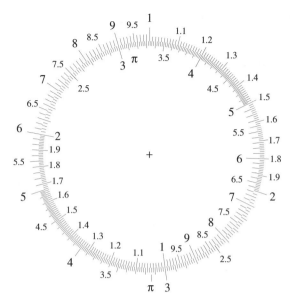

Figure 11.8. Circular rule set to multiply by three.

cylinder approximately 2 in long and $1\frac{1}{4}$ in in diameter. Details of the scale are shown in figure 11.10.

As an example of a circular layout, we can place one scale inside another. The two scales in figure 11.8 are set up in such a way that we can use them to multiply by three. The 1 on the inside is next to the 3 on the outside. Similarly, the 2 on the inside is next to the 6 on the outside. We can see that the advantage of a circular configuration is that we do not run out of scale when we come to 4 on the inside, which is next to the 1.2 on the outside. Since we have gone round the outside once this corresponds to 12, which is the correct answer. So, just as with a linear rule we still need to keep track of the decimal point.

11.5   Slide Rules for Special Purposes

As mentioned, most slide rules have gauge points that indicate constants to aid specific calculations, π being an example. A special slide rule is one that has whole scales inscribed in such a way that calculations relating to specific problems can be resolved quickly, simply and accurately. They have the advantage that a

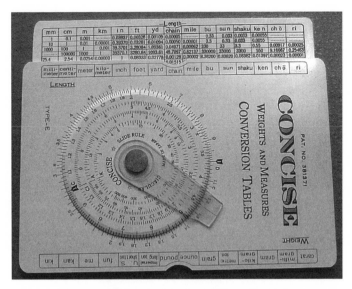

Figure 11.9. A circular slide rule with conversion tables, the 'Concise Type E'.

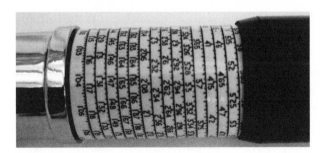

Figure 11.10. Detail of the helical scale on the Otis–King calculator.

problem can be resolved almost as quickly and with little more trouble than it takes to write the numbers down. In some circumstances this can be very important: when a flight heading has to be corrected to account for prevailing winds, for example. Furthermore, a specific scale can be more accurate than using a general scale. For example, for a calculation involving barometric pressure at or around sea level, numbers in the range 710–770 (millimetres of mercury) will be used. This range is compressed into less than 15 mm on a general-purpose 300 mm rule. A slide rule is also less bulky than a chart or graph.

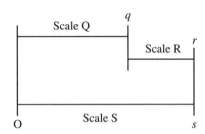

Figure 11.11. A general slide rule schematic.

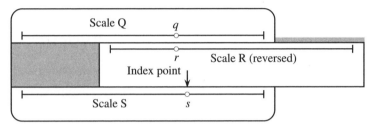

Figure 11.12. The usual scale layout on a special slide rule.

Of course, a special rule can be used for only one problem, or at most a very limited number. They are complicated and difficult to produce, especially when the scale markings are calculated by hand. This is a serious flaw and limited the range of applications in which special rules were used. Finally, suitable blank rules were difficult to obtain by non-specialist manufacturers. Despite these problems, many special rules were constructed and used. The following section briefly describes the theory and examines a particular case study: an oil tonnage calculator.

Special rules operate by adding lengths proportional to the values of the functions represented on their scales. For example, the scheme shown in figure 11.11 solves

$$Q + R - S = 0, \qquad\qquad (11.3)$$

where length $Q = f_1(q)$, length $R = f_2(r)$ and length $S = f_3(s)$. The scales are marked with numbers q, r and s. If f_1, f_2 and f_3 are all logarithms, then (11.2) becomes precisely (11.3), and we have the usual configuration for multiplication.

In order to prevent confusion over which scales to use, we assume that scale S contains the unknown, which is to say the

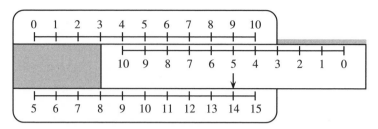

Figure 11.13. A slide rule for sums.

desired answer. It is then easiest to draw Q on the upper part of the stock and R contiguous with Q on the slider. The result is then read from scale S using a cursor or an index line on the slider. When constructing a special rule, we may not have a cursor available. If this is the case we have to use a fixed index point, rather than a moving cursor. To do this it is necessary to arrange the scales slightly differently. In particular, the scale R needs to be reversed. The r in figure 11.11 will then appear directly below the q, and the index point will be below scale R and directly above scale S: see figure 11.12. Apart from removing the need for a cursor, there are other advantages to this arrangement. Reversing the direction of scale R, and the flexibility we have in placing the index point, means that we are less likely to run out of scale S using this configuration. Taking a proportional scale for each of scales Q, R and S reduces (11.3) to

$$Q + R = S.$$

Now, taking S to be the unknown, we can devise a rule for adding numbers. This is shown in figure 11.13 and illustrates the reversed scale R. In this example, one immediately notices many different sums that total 14.

For an example of a reversed scale on a normal slide rule, look for the scale CI, which is usually drawn on the slider, often in red. This corresponds mathematically to $1/x$, and since $\log(1/x) = -\log(x)$, graphically this corresponds precisely to a reversed copy of the principal scale C. To multiply with this using the layout of scales illustrated in figure 11.12, imagine a copy of the usual D scale occupying the position of scale Q. Use the CI scale as scale R and take the 1 on the C scale as the index point.

The answer to normal multiplication of q and r can be read off directly using the index point and the D scale. What we have actually done here is divided by $1/x$, as represented on CI, which is of course multiplication.

For another example, using the standard slide rule, take scale Q as the scale A (the squared scale, x^2), R is the CI scale $(1/x)$, and take π as the index point on the C scale. Setting the slider for numbers on scales Q and R, the answer on the bottom scale is $s = \pi r \sqrt{q}$.

This is not the place to explain the intricate arrangements of scales in more complex rules or the techniques for efficiently laying these scales out. Such details can be found in the work of Hoelscher, Arnold and Pierce (1952).

11.6   The Magnameta Oil Tonnage Calculator

As an example of a special rule, consider the Magnameta oil tonnage calculator, which is a slide rule specifically constructed to deal with the problems relating to the loading of oil tankers. Imagine trying to balance the ship as oil of a specific density flows into tanks of various sizes positioned in various places around the ship. The trick is not to capsize the boat during this process. As the instruction book says,

> [i]t is intended primarily for use in handling loading problems, when the ability to estimate the weight per tank with speed and accuracy is of prime importance.

It continues as follows.

> The rule, which in spite of its size is neither heavy nor cumbersome, carried under the arm, is the answer to this problem. The official figure must under these circumstances be worked out later.

The rule is of standard design, as shown in figure 11.12, with a single slider and a cursor. The instrument has five scales denoted A–E. The primary scales A and E are marked on the stock and measure 783 mm in length. The other scales are marked on the slider and measure 381 mm. Scale A is the tank capacity, calibrated from 15 000 to 50 000 cubic feet. Scales B and C are international specific gravity values from 0.6 to 1.08. These are reversed logarithmic scales running from 33.292 to 59.972 with markings giving the specific gravities. The reversal of these scales is exactly as in figure 11.12. The D scale is marked in American petroleum industry (API) specific gravity values, and E is calibrated in tons for values between 450 and 1500 tons per tank.

The cursor has three lines, marked I, II and III from right to left. Lines II and III correspond to multiplication by 1.015 and 1.12 and the three lines allow calculations in long or short tons or in metric tonnes, respectively. The cursor is also fitted with a clamping device that holds it in one place relative to the slider. Loading problems always involve oil of a fixed specific gravity so being able to accurately fix the cursor has obvious advantages. Essentially the moving cursor becomes a fixed index point.

To perform a calculation, the cursor is fixed at a specific gravity and then the slider, with fixed cursor, is moved to a corresponding volume on the top scale A. The weight is read directly off scale E using the appropriate index point. This is exactly as described in section 11.5 above. The rule can be used for tanks calibrated in water tons and cubic metres (metric water tonnes) by using lines II and III on the cursor. Weight to volume calculations are straightforward using the reverse process. Lastly, API specific gravities can be converted to International specific gravity units by allowing the cursor to run freely and reading across scales C and D.

United Kingdom Patent 919063 was granted to the inventor Guy William Farrier Sangwin, the complete specification having been published on 20 February 1963 (Sangwin 1963). The calculations for the scale and the printing of the plates took three years to complete. The prototype rule, with scales printed onto plastic and laminated onto a wooden base, survives although the leftmost 150 mm of scales A and E have been badly damaged by water (the rule was rescued from a garage). This is

shown in plate 23. The introduction of cheap electronic calculators in the 1960s prevented this particular rule being produced
commercially.

11.7 Non-Logarithmic Slide Rules

Slide rules and logarithms are inextricably linked, so it may come
as a surprise to discover that to produce a slide rule capable of
multiplying two numbers it is not necessary to use logarithms at
all. We use this as another example of a special slide rule, since
the construction exactly follows that in section 11.5. The original
idea for this comes from Mills (1958).

Section 6.1 introduced the triangular numbers

$$1 + 2 + \cdots + n = \tfrac{1}{2}n(n + 1).$$

Instead of taking a whole number n, define for any positive number x the function

$$\Delta(x) := \tfrac{1}{2}x(x + 1).$$

Then, by elementary algebra, we have that

$$\Delta(a) - \Delta(a - b) + \Delta(b - 1)$$
$$= \tfrac{1}{2}(a(a + 1) - (a - b)(a - b + 1) + (b - 1)b)$$
$$= \tfrac{1}{2}(a^2 + a - (a^2 - 2ab + a + b^2 - b) + b^2 - b) = ab.$$
$$(11.4)$$

Given a and b, by calculating $\Delta(a) - \Delta(a - b) + \Delta(b - 1)$ we
may calculate ab (notice the similarity with (11.3)). It is on this
observation that the 'Nelson rule' relies. By drawing scales proportional to $\Delta(x)$ (instead of $\log(x)$ as in a logarithmic slide
rule) we may design a slide rule. The scales are drawn as follows:
the top scale is proportional to $\Delta(a)$, the top scale of the slider
is proportional to $\Delta(a - b)$, the bottom scale of the slider is
proportional to $\Delta(b - 1)$, and the bottom scale on the stock is a
simple proportional (equal division) scale.

Such a rule, set up to multiply 8 by 3, is shown in figure 11.14.
Set up the rule with the $a - b = 8 - 3 = 5$ on the top slider scale

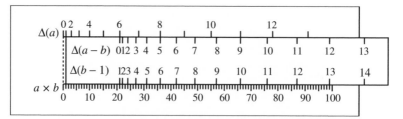

Figure 11.14. Nelson's rule.

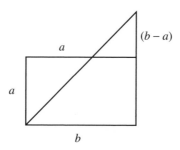

Figure 11.15. Geometrical interpretation
of improvements on Nelson's rule.

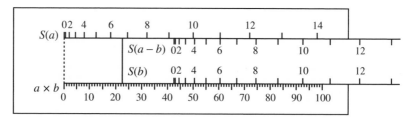

Figure 11.16. Improvements on Nelson's rule: 11×5.

adjacent to the $a = 8$ on the top scale. Look at the $b = 3$ on the
lower scale. This is above the answer, 24, on the proportional
scale. That Nelson's rule works for arbitrary positive numbers
and not simply whole numbers is obvious if looked at from the
point of view of the algebra in (11.4), rather then by viewing
$\Delta(x)$ in terms of triangular numbers. Nelson's rule may actually
be simplified, since algebra also confirms that

$$\tfrac{1}{2}a^2 - \tfrac{1}{2}(a - b)^2 + \tfrac{1}{2}b^2 = ab.$$

Comparing the areas of the triangles and rectangles shown in figure 11.15 confirms this geometrically. By using an almost identical construction, this time using the function

$$S(x) = \tfrac{1}{2}x^2$$

in place of $\Delta(x)$, we arrive at the slide rule shown in figure 11.16. This rule is set up to calculate $a \times b = 11 \times 5$ by setting $a = 11$ on the top scale adjacent to the $a - b = 6$ mark on the slide. The answer is read below $b = 5$ on the lower scale. The symmetry between a and b in the product $a \times b$ is more evident in this construction than in Nelson's. No doubt other relationships along similar lines are possible.

11.8 Nomograms

A related topic is that of *nomographs*, or alignment charts. These were frequently used for routine engineering calculations and the example we take is typical of the many that were in common use. It relates the volumetric flow rate of a gas to the nozzle diameter and the supply pressure. It consists of equispaced vertical logarithmic scales graduated in the accepted units: pressure in inches of water and flow rate in cubic feet per hour. The scale on the nozzle diameter axis looks more like an irregular arithmetic scale of numbers with no units on one side with a size in inches against each number on the other. Drills are available only in standard sizes and the number refers to those drills widely used for all sorts of purposes and still used by some model engineers. It would be pointless and prohibitively expensive to have to make a special drill to a calculated diameter for every nozzle.

As the name alignment chart suggests, it is used by placing a straight edge between the variables to find the value of the third. This is a quick and virtually foolproof way of doing a particular calculation.

An early example of a nomogram is a method devised by Sir Isaac Newton for approximating roots of polynomial equations using slide rules. It first appears in Newton's 'Waste book' (see

Whiteside 1961) around 1665. The method appears to have then been explained to John Collins in a letter of 1672. Collins was interested in finding the volume of liquid in a partially full barrel as a function of depth.

> Wherefore since I understand your designe is to get a rule for guageing vessells, this Problem having so bad success for yt end I shall in its stead present you with this following expedient.
>
> Turnbull (1959–61, volume 2, p. 229)

This was an important issue in gauging, i.e. the calculation of various taxes on liquids. Interestingly, slide rules were particularly popular amongst customs officials, for whom they provided an instant volume calculation, by visual inspection, and therefore instantly settled the question of how much tax or duty was due. It is reported in Whiteside (1961) that this method was incorporated by Collins into a general method for solving cubic equations. Collins then wrote *In Answer to Mon^r Leibnitz's Letter about Solving a Cubick Æquation by Plaine Geometry*. This 'answer' was passed to Oldenburg, who, in the following letter dated 24 June 1675 (quoted in Cajori (1909, p. 21) but also in Horsley (1782, VI:520)), communicated the ideas to Leibnitz.

> Mr. Newton, with the help of logarithms graduated upon scales by placing them parallel at equal distances or with the help of concentric circles graduated in the same way, finds the roots of equations. Three rules suffice for cubics, four for biquadratics. In the arrangement of these rulers, all the respective coefficients lie in the same straight line. From a point of which line, as far removed from the first rule as the graduated scales are from one another, in turn, a straight line is drawn over them, so as to agree with the conditions conforming with the nature of the equations; in one of these rules is given the pure power of the required root.

The theory is described briefly in Cajori (1909), and reference is made to Stone (1743) in which 'Newton's mode of solving equations mechanically is explained more fully and with some restrictions, rendering the process more practical'. Both the linear and circular versions are discussed in more detail in Sangwin (2002). We have used this technique, but it is not easy.

11.9 Oughtred and Delamain's Views on Education

We digress slightly in this section and comment on Oughtred and Delamain's views on education. We know something of these because of the priority dispute for the invention of the slide rule (discussed in section 11.2). Delamain, originally a joiner but later a 'teacher of practical mathematics', was eventually to become tutor to King Charles I. He was also employed by the Crown to make a number of mathematical instruments. These included sundials, perhaps even the silver sundial worn by Charles I at his execution.

Oughtred was a much-respected teacher of mathematics. A clergyman by profession, his skill as a teacher was well known and was sought out by his contemporaries. There is little doubt that Oughtred was the more competent mathematician. Isaac Newton, whose annotated copy of Oughtred's book *Clavis Mathematicæ* ('Key to Mathematics') survives to this day, replied on 25 May 1694 to a request for a recommendation on a proposed mathematics course thus:

> And now I have told you my opinion in these things, I will give you Mr Oughtred's, a Man whose judgment (if any man's) may be safely relied upon.
>
> Turnbull (1959–61, volume 3, p. 364)

Oughtred emphasizes that mastery of the 'art', or theory, will lead naturally to the mastery of the practice, and in particular that instruments should not be introduced too early. The dedication to Sir Kenelm Digby contained in *Circles* was written by William Forster, a teacher of mathematics in London who was also a friend and pupil of Oughtred, who undertook to translate and publish Oughtred's Latin notes. Of Oughtred's attitude, Forster says the following.

> He answered, That the true way of Art is not by Instruments, but by Demonstration: and that it is a preposterous

course of vulgar Teachers, to begin with Instruments and not with the Sciences, and so instead of Artists to make their Schollers only doers of tricks, and as it were Juglers: to the despite of Art, losse of presious time, and betraying of willing and industrious wits unto ignorance and idlenesse. That the use of Instruments is indeed excellent, if a man be an Artist; but contemptible, being set and opposed to Art. And lastly, that he meant to commend to me the skill of Instruments, but first he would have me well instructed in the Sciences.

In the *Apologie* Oughtred emphasizes that a grasp of theory will lead naturally to the mastery of practice. This is entirely consistent with his experience: Oughtred had invented many mathematical instruments in addition to calculating devices (Turner 1973). Oughtred de-emphasizes the benefits of instrument *use* to such an extent that he claims not to regard his own invention highly. As he recounts in the *Apologie*:

The Instruments I doe not value or weigh one single penny. If I had been ambitious of praise, or had thought them (or better than they) worthy, at which to have taken my life, out of my secure and quiet obscuritie, to mount up into glory, and the knowledge of men: I could have done it many years before this pretender know any thing at all in these faculties.

This is, perhaps, somewhat contradictory on two counts: he felt strongly enough to pen the 32-page *Apologie* defending his position and he recounts using these instruments profitably in his own work. Furthermore, Oughtred was sufficiently content with his 'quiet obscuritie' in the period 1631–33 to also publish *Clavis*, *Horizontall Dyall*, *Gauging Line*, and an edition of his *Mathematical Recreations*. Given that one of his most enduring legacies to science is the slide rule, it is indeed a shame he did not hold it in high regard. Of course, it must be taken in the context of the dispute with Delamain, which by this time had become heated. The phrase 'vulger teacher' in Forster's introduction to *Circles* could only apply to Delamain, to which he unsurprisingly took strong objection. Indeed, Oughtred described *Grammelogia* as follows.

I borrowed and perused that worthless Pamphlet, and in
reading it (I beshrew him for making me cast away so much
of that little time remaining to my declining years) I met with
such a patchery and confusion of disjoynted stuffe, that I
was stricken with a new wonder that any man should be so
simple, as to shame himself to the world with such a hotch
potch.

Oughtred goes further, clearly stating that a mastery of the
principles involved inevitably leads to an ability to *reinvent* them
for one's self:

But *Delamain* was already corrupted with doing upon Instruments, and quite lost from ever being made an Artist: I
suffered not *William Forster* for some time to so much as
to speak of any Instrument,... And this my restraint from
such pleasing avocations, and holding him to the strictness
of precept, brought forth this fruit, that in short time, even
by his owne skill, he could not onely use any Instrument he
should see, but also was able to delineate the like, and devise
others.

Richard Delamain, on the other hand, took a strikingly different view of the theory before practice debate:

But, for none to know the use of a *mathematical instrument*,
except he who knows the cause of its operation is somewhat too strict, which would keep man from affecting the
Art, which of themselves are ready enough everywhere, to
conceive more harshly of the difficultie, then it deserves,
because they see nothing but obscure propositions.

Furthermore, Delamain believed that theory and practice should
coexist and in doing so would strengthen each other:

The beginning of a *mans doctrinal precepts*, and these precepts may be conceived all along in its use: and are so farre
from being excluded, that they doe necessarily *concomitate*
and are contained therein: the *practike* being better understood by the *doctrinall part*, and this better explained by the
Instrumentall, making precepts obvious unto sense, and the

> *Theory* going along with the *Instrument*, better informing and inlighting the understanding etc.

Delamain also clearly acknowledges a distinction between different groups of students:

> All are not of like disposition, neither all (as was sayd before) propose the same end, some resolve to *wade*, others to put a *finger* in onley, or wet a *hand*: now thus to tye them to an obscure and *Theoricall* forme of teaching, is to crop their hope, even in the very bud.

He continues with a plea which is strikingly familiar:

> And me thinkes in this queasy age, all *helpes* may bee used to procure a *stomacke*, all *bates* and invitations to the declining studie of so noble a *Science*, rather then by rigid Method and generall *Lawes* to scarre men away.

Chapter 12

ALL A MATTER OF BALANCE

What is it indeed that gives us the feeling of elegance in a solution, in a demonstration? It is the harmony of the diverse parts, their symmetry, their happy balance; in a word it is all that introduces order, all that gives unity, that permits us to see clearly and to comprehend at once both the ensemble and the details.

Henri Poincaré

This chapter starts our discussion of *balance*. We begin by balancing identical objects one on top of the other in such a way that they do not fall over. Then we move on to other problems in a similar vein. Rather than asking how to make something that balances, these ask how we can make a solid object which will not tip over when placed on a table.

12.1 Stacking Up

To begin the chapter we propose another experiment: for this you will need a large supply of identical block-like solid objects— dominoes are ideal and CD cases also work very well. You should try to stack these up one on top of another in such a way that the resulting tower leans over without collapsing. The question is the following.

> If dominoes are stacked with only one touching the table, what is the greatest horizontal distance we can cover without tipping?

As we shall see, the solution to this apparently innocuous problem will take us on a long and productive flight of fancy.

We have restricted ourselves to only allowing one on top of the other, since by constructing a cantilever structure such as

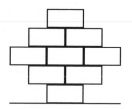

we can cover any horizontal distance we choose! Clearly this solution to the problem is not very interesting, and what we are really interested in solving is the problem of the maximum distance we can cover using a *leaning stack*. The stack must balance with only one domino per level. So how far horizontally can we get a tower of dominoes, one on top of the other, to lean? There are to be no counterbalancing or cantilever-type structures here!

Naturally, taking large dominoes we could obviously cover a large distance, so we think of the distance covered in units relative to the length of a single domino. Can we cover a distance twice the length? Three times? Hence we assume that the length of the block is 2 arbitrary units and that the block has mass 1: see figure 12.1. For now, the height and width of the dominoes will play no role. Let us assume that we have a leaning stack of n dominoes one on top of the other which leans to the right. We define c_n to be the horizontal distance from the leftmost domino to the centre of mass of the stack.

To enlarge the stack we add another domino without destroying it, but this poses a problem. If we add a domino to the top the stack could break at any point, since none of the dominoes are glued together. To do this successfully we would need to check that at no point does the stack topple over.

A simpler way is to proceed by induction: placing the existing stack *on top* of the new domino making sure we do this in such a way that the centre of mass of the existing stack is above the new domino (see figure 12.2). So our strategy is this: at each stage we place the existing (balancing) stack on top of a new domino a

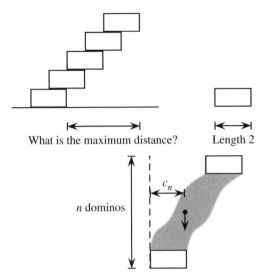

Figure 12.1. A leaning stack.

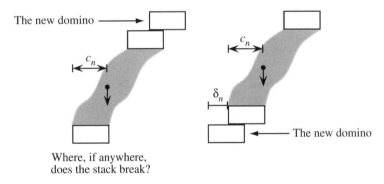

Where, if anywhere,
does the stack break?

Figure 12.2. Adding a domino at the top and bottom of the stack.

distance δ_n from the left of the domino. There will clearly be no toppling if

$$\delta_n + c_n \leqslant 2 \quad \text{for all } n. \tag{12.1}$$

That is to say, we do not displace the top stack so far that the displacement plus the distance of the centre of mass from the left pushes the centre of mass over the edge of the bottom domino. The new centre of mass of the whole stack of $n + 1$ dominoes will be c_{n+1} from the left of the bottom domino, where

$$c_{n+1} = \frac{(\delta_n + c_n)n + 1}{n + 1} \quad \text{with } c_1 = 1 \text{ (one domino).}$$

As a check let us calculate what happens if we build a vertical tower with no displacement at all—that is to say, $\delta_n = 0$ for all n. Then, from our relation above,

$$c_{n+1} = \frac{c_n n + 1}{n + 1}.$$

Given that $c_1 = 1$ we see, by induction, that $c_n = 1$ for all n also. So for a vertical stack, the centre of mass is always in the middle, as expected.

Let us push the dominoes as far as is theoretically possible. By this we mean we move the centre of mass so that it is right on the edge of the bottom domino. For two dominoes this means $\delta_1 = 1$ and we have the following configuration:

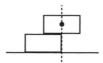

Remember that a domino will only topple if there is a positive rotational force generated by the centre of mass being *over the edge*. An arbitrary small clockwise tip will topple the stack in this situation, but for the moment, the fact that it balances in theory is sufficient.

To continue this we would need, this time, to choose

$$\delta_n := 2 - c_n. \tag{12.2}$$

If we do this and derive our sequence for c_n, we find that

$$c_{n+1} = \frac{(\delta_n + c_n)n + 1}{n + 1} = \frac{(2 - c_n + c_n)n + 1}{n + 1} = \frac{2n + 1}{n + 1}.$$

Using this in (12.2) gives

$$\delta_{n+1} = 2 - c_{n+1} = \frac{1}{n + 1} \quad \text{with } \delta_1 = 1.$$

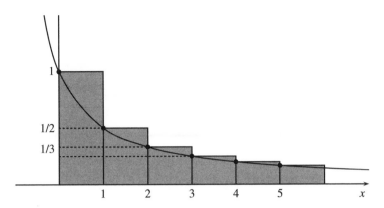

Figure 12.3. Comparing the series $1/n$ with the function $1/(1 + x)$.

That is to say,

$$\delta_n = \frac{1}{n} \quad \text{for all } n.$$

So the question becomes, what is the value of

$$1 + \frac{1}{2} + \frac{1}{3} + \frac{1}{4} + \cdots + \frac{1}{N} \qquad (12.3)$$

for large N? Ideally we would like to let $N \to \infty$, as we did in equation (6.3). The fact that each term in the infinite sum gets smaller, and ultimately converges to zero, gives us some hope that the infinite sum might total to a finite value. Actually, this hope is quite misplaced and what we have is a particularly famous series, known as the *harmonic series*. In the next section we will show that this series diverges. That is to say, it is possible to make the sum (12.3) as large as one would wish. In terms of the domino problem: we can chose displacements δ_n so that (i) the stack does not topple over and (ii) we can produce an arbitrarily large horizontal displacement. Bizarre indeed!

12.2 The Divergence of the Harmonic Series

The following medieval proof that the harmonic series diverges was discovered and published by Nicole d'Oresme (1323-82) around 1350 and relies on grouping the terms as follows:

$$1 + \tfrac{1}{2} + \tfrac{1}{3} + \tfrac{1}{4} + \tfrac{1}{5} + \tfrac{1}{6} + \tfrac{1}{7} + \tfrac{1}{8} + \tfrac{1}{9} + \tfrac{1}{10} + \cdots$$

$$= 1 + \tfrac{1}{2} + \underbrace{(\tfrac{1}{3} + \tfrac{1}{4})}_{>2 \times \frac{1}{4}} + \underbrace{(\tfrac{1}{5} + \tfrac{1}{6} + \tfrac{1}{7} + \tfrac{1}{8})}_{>4 \times \frac{1}{8}}$$

$$+ \underbrace{(\tfrac{1}{9} + \tfrac{1}{10} + \tfrac{1}{11} + \tfrac{1}{12} + \tfrac{1}{13} + \tfrac{1}{14} + \tfrac{1}{15} + \tfrac{1}{16})}_{>8 \times \frac{1}{16}} + \cdots$$

$$> 1 + \tfrac{1}{2} + \tfrac{1}{2} + \tfrac{1}{2} + \tfrac{1}{2} + \cdots .$$

It follows that the series can be made arbitrarily large, although it diverges quite slowly.

Calculus, although unnecessary for our story, can also be used to provide an explicit estimate of the rate of growth by comparing the graph of a function with the terms of the series. In particular we compare terms in the series with the area under the graph of the function $1/(1+x)$. Figure 12.3 should be enough to convince most people that

$$1 + \frac{1}{2} + \frac{1}{3} + \frac{1}{4} + \cdots + \frac{1}{N} > \int_0^N \frac{1}{1+x} \, dx.$$

We need to recall general calculus theorems and the fact that

$$\int_1^t \frac{1}{x} \, dx = \ln(t), \tag{12.4}$$

where the logarithm is the natural logarithm, i.e. a logarithm to the base e. In fact, in mathematics, (12.4) is usually taken to *define* the logarithm function. From this starting point logarithms to other bases can be defined and the rules of logarithms, such as (11.2), derived.

The integral can be solved giving

$$1 + \frac{1}{2} + \frac{1}{3} + \frac{1}{4} + \cdots + \frac{1}{N} > \ln(1+N).$$

Now, the function $\ln(1+N)$ is unbounded so we can make the left-hand side of the above expression as large as we please by

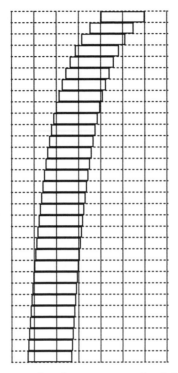

Figure 12.4. The diverging stack of dominoes.

taking sufficiently large N. A similar argument comparing the series to the function $1/x$ shows that

$$1 + \ln(N) > 1 + \frac{1}{2} + \cdots + \frac{1}{N} > \ln(1 + N), \qquad (12.5)$$

so we can estimate how fast the series diverges.

12.3 Building the Stack of Dominos

It is perhaps inevitable that having realized that there is no theoretical limit to the potential horizontal distance that the stack of dominoes can cover we might like to see what it looks like. Can we really construct such a stack in practice and, if so, how would we do this?

Recall that our scheme for choosing δ_n was only one of many possibilities. If we choose δ_n to be as big as possible at each stage, then any small inaccuracy in our construction will doom our efforts to failure. Instead, a much safer practical approach is to choose δ_n to be consistently smaller than this, perhaps taking $\delta_n := 1/(2n)$. Instead of this simple scheme we choose δ_n to be

$$\delta_n := \tfrac{1}{2}(2 - c_n), \tag{12.6}$$

which is half the maximum possible distance to the toppling point. We call this a '50% stack'.

This is more conservative and the resulting stack is no longer exactly on the point of toppling. Taking this scheme, what is δ_n? And more importantly, what is the total horizontal distance given by $\delta_1 + \delta_2 + \cdots + \delta_N$? Indeed, will we still be able to achieve an arbitrarily large displacement?

First let us calculate the c_ns explicitly under this scheme:

$$c_{n+1} = \frac{(\delta_n + c_n)n + 1}{n + 1} = \frac{(2 + c_n)n + 2}{2(n + 1)} = \frac{nc_n}{2(n + 1)} + 1.$$

Now we set $b_n := 2 - c_n$ and see what happens to the b_ns:

$$b_{n+1} = 2 - c_{n+1} = 1 - \frac{nc_n}{2(n + 1)}$$

$$= 1 - \frac{n(2 - b_n)}{2(n + 1)} = \left(1 - \frac{n}{n + 1}\right) + \frac{b_n}{2}\frac{n}{n + 1} \geqslant \frac{1}{n + 1}.$$

So, by induction,

$$b_n \geqslant \frac{1}{n}.$$

However,

$$\delta_n := \frac{2 - c_n}{2} = \frac{b_n}{2} \geqslant \frac{1}{2n}.$$

So we see that the total displacement is $\sum_n \delta_n$ and if we take N dominoes in the stack this exceeds

$$\sum_{n=1}^{N} \frac{1}{2n} = \frac{1}{2}\sum_{n=1}^{N} \frac{1}{n}.$$

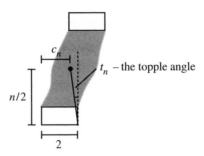

Figure 12.5. Toppling the stack.

What we really want is an explicit recurrence relation for δ_n and we can calculate this as follows:

$$2\delta_{n+1} = 2 - c_{n+1} = 2 - \frac{nc_n}{2(n+1)} - 1$$
$$= \frac{2n + 2 - 2n(1 - \delta_n)}{2(n+1)} = \frac{1 + n\delta_n}{n+1}.$$

Thus

$$\delta_{n+1} = \frac{1 + n\delta_n}{2(n+1)} \quad \text{with } \delta_1 := \frac{1}{2}.$$

Unfortunately this recurrence relation does not seem to have a simple closed-form solution.

Instead, we calculate successive values of δ_n numerically: these are shown in table 12.1. Notice also how close the value of c_n gets to the critical value of 2, which is the value above which the stack will overbalance the bottom domino. If we use this recurrence relation and take blocks of length 2, we can construct the leaning tower shown in figure 12.4, where, for the purposes of illustration, the height is $\frac{1}{2}$.

Given that c_n approaches the critical value of 2 so closely, we examine the *angle* through which we would need to tip the whole stack, about the bottom corner, to topple it over.

The centre of mass of an object is fixed within an object regardless of how it is tipped or rotated. For our stack of dominoes this is c_n from the left-hand edge of the bottom block, and therefore $2 - c_n$ from the right-hand edge—this is illustrated in figure 12.5. We know that the blocks have length 2 but up to this point their

Table 12.1. The diverging stack of dominoes.

n	c_n	δ_n	t_n
1	1.000	0.500	75.964
2	1.250	0.375	56.310
3	1.417	0.292	37.875
4	1.531	0.234	25.115
5	1.613	0.194	17.223
6	1.672	0.164	12.339
7	1.717	0.142	9.201
8	1.751	0.125	7.097
9	1.778	0.111	5.630
10	1.800	0.100	4.569
11	1.818	0.091	3.781
12	1.833	0.083	3.179
13	1.846	0.077	2.710
14	1.857	0.071	2.337
15	1.867	0.067	2.036
16	1.875	0.062	1.790
17	1.882	0.059	1.586
18	1.889	0.056	1.414
19	1.895	0.053	1.270
20	1.900	0.050	1.146
21	1.905	0.048	1.039
22	1.909	0.045	0.947
23	1.913	0.043	0.866
24	1.917	0.042	0.796
25	1.920	0.040	0.733

height has played no role. We take the height of a block to be h so that the centre of mass is $\frac{1}{2}hn$ from the bottom of the stack. Thus we see that the angle of rotation t_n needed to topple the stack is

$$t_n = \arctan\left(\frac{4 - 2c_n}{hn}\right).$$

Values for t_n when $h = \frac{1}{2}$ (so the length of the block is four times the height) are also given in table 12.1 (in degrees).

This also illustrates the difference between stability and instability. Our scheme (12.6) gave rise to a stable stack—there exists a small but positive displacement through which you can tip the stack without toppling it. As Eisner (1959) comments:

> To prove this result 'physically', a fellow graduate student and I stacked bound volumes of *The Physical Review* one evening, until an astonishingly large offset was obtained, and left them to be discovered the next morning by a startled physics librarian.

12.4   The Leaning Pencil and Reaching the Stars

Your intuition has probably rejected this result, and even if you accept the mathematics you may still harbour some doubts. While the result holds, your intuition is probably not that poor, and we shall now give a practical example to reassure you.

Suppose we wish to span a distance of 3 m with CD cases, which actually balance rather well. A CD case measures about 14 cm $\times$ 12 cm $\times$ 1 cm. So we want

$$1 + \frac{1}{2} + \frac{1}{3} + \frac{1}{4} + \cdots + \frac{1}{N} = 3 \text{ m},$$

where the left-hand side is in units of 'half a CD case', which is about 7 cm. So, including the units we need,

$$1 + \frac{1}{2} + \frac{1}{3} + \frac{1}{4} + \cdots + \frac{1}{N} = \frac{300}{7} = 42.857.$$

By the approximation (12.5) this gives

$$N \approx e^{42.857\cdots} \approx 4.1 \times 10^{18}.$$

Now, the thickness of a CD case was taken to be 1 cm, so the height of the stack of CD cases needed to cover a horizontal distance of 3 m would be

$$4.1 \times 10^{16} \text{ m} = 4.3 \text{ light years}$$

(one light year is 9.46×10^{15} m). If you tried to build it the gravitational attraction between the CD cases would probably

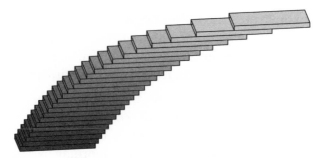

Figure 12.6. An unlikely looking, but nevertheless balancing, stack!

be much more significant than the Earth's gravitational pull on the CD cases, making the problem of 'balance' more tricky.

While CD cases, or even a pack of playing cards, are a good choice for a temporary model, it is also satisfying to glue wooden blocks in place. While this is not necessary from a theoretical point of view it does allow you to take the time to arrange the blocks into an impressive stack for illustrative purposes. The model shown in plate 24 was made from a length of ramin strip measuring 1 in × $\frac{1}{4}$ in (25 mm × 6.25 mm), which was obtained from a DIY shop. The wood takes a good finish and is hard enough for the complete model to withstand normal handling. The first stage in making a model such as this is to decide on the length of each tile. One unit equal to 25 mm or 1 in is a convenient scale. Then prepare a table of displacements δ_n, and again it is convenient to calculate them from

$$\delta_n = \frac{1}{n}(1 - (1 - f)^n),$$

where f is the fractional approach to the balance point, so that when $f = 0$ a vertical tower results and when $f = 1$ the displacements become the harmonic series. Of course, there are other possibilities for δ_n provided each is less than or equal to $1/n$. This is one instance where we found it easier to work with imperial measurements for δ_n and give the displacements directly in fractions of an inch.

Precision at every stage of construction is vital, otherwise the model will not meet expectations and will most probably fall over. With this in mind our strip was cut into lengths of 2 in, measured with Vernier callipers. It is essential to make a saw guide and use a fine-tooth razor saw. You can expect to waste several lengths before you are sure that the length is true, and even then it is worth checking each length as it is sawn off to make certain nothing has shifted. As with many jobs in a workshop it may well take more time to make the sawing fixture than to cut the actual pieces of wood. Any ragged ends left after sawing can be sanded off, and even here some precautions are necessary. Glue the paper to a flat piece of wood and only sand along the length of the tile before cleaning the end faces. Those with access to a small saw bench have a considerable advantage.

To build the model itself a top-down construction is needed. Start with the top pair of tiles and with a thin mixture of woodworking glue fix them together using the Vernier callipers to get the displacements exactly right. It is best to do this on a flat surface using a square to check they are in line on their sides. Cross out this step on your pre-prepared table of displacements and proceed to the next tile. After four or five tiles put the model on its side and sight along the upper corners to make sure everything lies on a smooth curve.

The challenge of this model is to see how tall you can make it, and as you can see from plate 25, we have managed thirty tiles with $f = 50\%$. Our efforts with $f = 70\%$ and $f = 90\%$ are more modest. However tall you can reach, it is always satisfying when the top tile overhangs the base completely, as somehow this does not 'look right'. Furthermore, we might be able, if we took very small blocks, to sand it down to give a smooth finish. We would then have a bent pencil-like wooden shape that potentially reaches as far as we like across the table but balances on its base.

12.5 Spiralling Out of Control

All the stacks described so far have been linear, in the sense that the longitudinal centre lines of individual tiles all lie on the same straight line in plan view. This is not a necessary condition for stability, however, as we shall show. Suppose the top n tiles

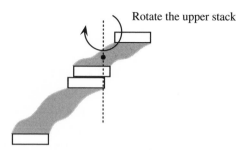

Figure 12.7. Rotating part of the stack in the horizontal plane.

Figure 12.8. Even this balances!

are removed from a linear stack: the centre of mass then lies at a distance c_n from the left-hand edge of the bottom tile. The whole stack is stable if this load of n tiles is placed on the $(n + 1)$th tile with its centre of mass at $(\delta_n + c_n)$. At this stage nothing has been said about the particular arrangement of these top n tiles, only that the position of their centre of mass must be at c_n, according to the protocol set out at the start of this chapter. The top part of the stack could be turned through 180°, as shown in figure 12.7, or indeed through any angle without loss of stability, as illustrated by plate 26.

In principle a stack could be pivoted to swing at more than one level and through any angle, as in figure 12.8. There is, however, a practical problem. Pivot holes have to be drilled and as

n increases their position becomes too close to the end of the tile beneath to be practicable. The purist would always make the pins of the same material as the tiles. In the model we used 1 mm plastic rod. The pivot pins bear no load except when the stacks are rotated.

We finish with a cautionary note. Our original intention was simply to build the tall stack shown in figure 12.4 using the calculated values in table 12.1. Once we had finished it, though, we found that we wanted to build others. So buy plenty of wood strips and retain your sawing jig for future use in case you too are tempted! Indeed, the recent article of Hall (2005) contains other possibilities we have yet to fully exploit.

12.6   Escaping from Danger

One reason why the harmonic series is so important in mathematics is that it illustrates the following result. Just because the terms in a sum converge to zero, the sum itself does not necessarily have a finite value. That is to say, while the sequence $1, \frac{1}{2}, \frac{1}{3}, \ldots$ converges to zero, the sum $1 + \frac{1}{2} + \frac{1}{3} + \cdots$ is infinite. The following problem, posed by Alan Slomson, is one example of many and exploits the same mathematics that helped us stack dominoes.

> Imagine a worm sitting on a 100 m long elastic tightrope. This worm moves at a constant speed of 1 m/s and merrily slithers along the rope, hoping to reach the other end. Yes, it's a fast worm.
>
> Unfortunately for the worm, a malevolent beast keeps stretching the rope instantaneously by 100 m at the end of every second. Note, the worm doesn't stretch since it is a point-size worm. So, at the end of the first second the rope is now 200 m long. At the end of the second second, the worm has travelled another 1 m and the rope is now 300 m long.
>
> When does the worm get to the end of the rope?

The situation appears hopeless, because the worm only moves 1 m before the rope becomes 100 m longer. However, remember that the rope *behind* the worm stretches as well as the rope in front. One way to think about this is to decide *what fraction* of the rope the worm has crossed in any given second. Thus in the first second it moves across one one-hundredth of the rope, during the second second, one two-hundredth, and in general it crosses a fraction $1/100N$ of the rope during the Nth second. The question becomes, when is

$$\frac{1}{100}\left(1 + \frac{1}{2} + \frac{1}{3} + \cdots + \frac{1}{N}\right) > 1.$$

That is, when has the worm passed the end? Of course, we know that the harmonic series diverges and that

$$1 + \frac{1}{2} + \frac{1}{3} + \frac{1}{4} + \cdots + \frac{1}{N} > \ln(1 + N).$$

If $\ln(1 + N) > 100$, we will be done. Now, inverting each side gives $N + 1 > e^{100} \approx 2.7 \times 10^{43}$ seconds. The short answer to the original question is 'not soon', but the worm *will* get to the end eventually.

12.7 Leaning Both Ways!

Despite the fact that we have been stacking up solid blocks, our problem was a fundamentally two-dimensional one. Originally we only stacked upwards and displaced the tower in one horizontal direction. While we rotated part of the stack around a vertical axis in section 12.4, there is a more interesting way to exploit the third dimension. This is to displace each block a distance δ_n in both of the horizontal directions simultaneously. If we take flat and square objects, rather then rectangular dominoes, we can construct a most implausible tower, such as that shown in figure 12.9.

What is so remarkable about this is that the top block has been displaced by (just under) a half in both horizontal directions. Hence, about three quarters of it appears to be unsupported. Nevertheless, this does actually work and a practical trial with CD cases is a pleasure not to be missed.

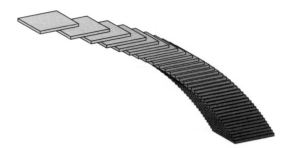

Figure 12.9. Leaning in two directions.

12.8 Self-Righting Stacks

In this section we shall assume that we have built a stack, but that this time all the dominoes are glued together. Stacks built according to the protocol just described can be said to be bistable. We are not concerned here with tipping sideways or backwards, only in the direction of the overhang. In their upright positions they are stable, but capable of withstanding only small disturbances before tipping over to a completely stable equilibrium. We now turn our attention to uni-stable stacks. That is to say, those for which the tipped-over position is *unstable* so that when the force that tipped them is removed they recover their original upright position. In other words, self-righting stacks. Almost inevitably these stacks are not as tall and need fewer tiles, but nevertheless they are interesting to make. It should be added that they still look as if they might tip quite easily.

The geometry needed to satisfy this new criterion is best illustrated by two examples shown in the sketch in figure 12.10. The stack with four tiles is uni-stable because when it is tipped over its centre of mass still lies to the left of the vertical y-axis, with the right-hand edge of the bottom tile at the origin O. When the force that tipped it is removed, the turning moment about the origin is anticlockwise and so the initial upright position is restored. When the five-tile stack is tipped, its centre of mass passes through the y-axis and since there is now no restoring moment it stays in that position in stable equilibrium: see plate 27.

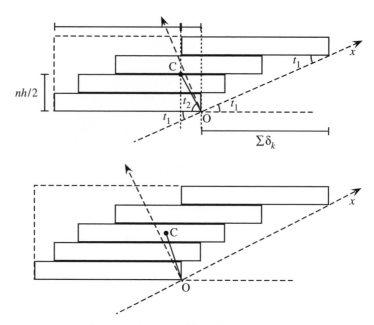

Figure 12.10. Self-righting stacks.

The two crucial angles are

$$\tan(t_1) = \frac{(n-1)h}{\sum_{k=1}^{n-1} \delta_k} \quad \text{and} \quad \tan(t_2) = \frac{nh}{2(2-c_n)}.$$

For the stack to be self-righting we require $t_1 + t_2 < \frac{1}{2}\pi$, or 90°. This inequality looks like a typical engineering design problem to be satisfied with the two variables n and h and the δs. Replacing the inequalities by

$$\varepsilon = \tfrac{1}{2}\pi - t_1 - t_2$$

still leaves a practical problem of the minimum value of ε that will ensure that the stack is self-righting without any external encouragement. All we can add is that the better the workmanship the smaller this angle can be: there is room here for the craftsman to experiment.

We can suggest further objectives: maximizing n or the total height nh, for example, or obtaining the greatest overhang or even the maximum value of h. All this strongly suggests a series of computer trials before making any model stacks. On a practical note, if the computed value of h is not a convenient size in terms of commercially available strip wood, the easiest solution is to use the nearest size and scale the unit length accordingly. As a final remark, we found that for a deck of playing cards the measured value is $h \approx 0.007$. Not a tall stack, perhaps, but plenty of overhang.

12.9   Two-Tip Polyhedra

So far in this chapter we have considered the stability of an object when one flat face is placed on a table. Now we turn our attention to polyhedra and ask which faces are stable faces. That is to say, when a polyhedron is placed on a table on a particular face, does it tip over or not? The criterion used to determine whether a polyhedron will tip or not is identical to that used before. The centre of mass must be directly above the base when the shape is placed on the table. Equivalently, a face is stable if, and only if, the orthogonal projection of the centre of mass onto the face lies inside the face or on the edge. Throughout this discussion we shall assume that the object has flat (planar) faces, is made of a material which is uniform in density and is placed on a horizontal table.

Given a homogeneous polyhedron, at least one of the faces must be stable. If this were not the case, then none of the faces would be, and no matter how the object were placed on the table it would tip over onto another face, and then keep moving. We can appeal to the impossibility of perpetual motion to persuade you that this will not occur. At least one face is stable.

The simplest polyhedron has four sides: the tetrahedron. A regular tetrahedron, the familiar triangular pyramid, is stable on all its faces. In general, an irregular tetrahedron is always stable on the face that is closest to its centre of mass, since it can have no lower potential energy. In fact, a tetrahedron must have at least *two* stable faces.

Can we make one? In fact it is possible to construct a *two-tip tetrahedron*, which when placed lying on one of its faces will tip over to another face and then tip again, finally coming to rest on a third face. The following solution is given by Heppes (1967).

We assume that the tetrahedron is sitting on the table and label the four vertices as A, B, C and D as follows:

$$A = (-7, -8(1 - \varepsilon), 0),$$
$$B = (-1, 0, 0),$$
$$C = (1, 0, 0),$$
$$D = (7, 8, 8),$$

where $0 < \varepsilon < 1$. Notice that it is the face ABC that sits on the table, with $z = 0$. The coordinates of the centre of mass, G, are given by

$$G = \tfrac{1}{4}(A + B + C + D) = (0, 2\varepsilon, 2),$$

so the orthogonal projection of the centre of mass onto the table has (x, y) coordinates $(0, 2\varepsilon)$, which lies outside the triangle ABC. Hence, the tetrahedron will tip on the edge BC, which lies on the x-axis. This involves a rotation of $45°$, and the new coordinates are

$$A' = (-7, -4\sqrt{2}(1 - \varepsilon), 4\sqrt{2}(1 - \varepsilon)),$$
$$B' = (-1, 0, 0),$$
$$C' = (1, 0, 0),$$
$$D' = (7, 4\sqrt{2}, 0).$$

Again, $G' = (0, \sqrt{2}(1 - \varepsilon), 0)$ lies outside the triangle B'C'D' and so the tetrahedron must tip over once more. The freedom to choose ε gives many such two-tip tetrahedra—an example is shown in plate 28.

12.10 Uni-Stable Polyhedra

For the tetrahedron, at least two faces are stable and there exist 'two-tip' tetraheda for which only two faces are stable. What about other polyhedra? We know all polyhedra have at least one stable face, but can we find an example of a convex polyhedron

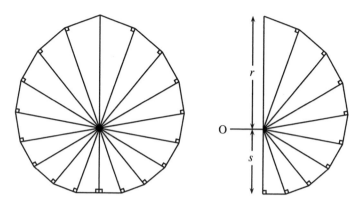

Figure 12.11. A cross-section of a uni-stable polyhedron.

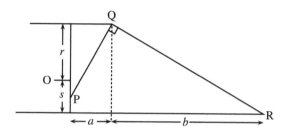

Figure 12.12. Truncating the prism.

that will rest in a stable position on *only one* face? The implication of this would be that no matter how the object were placed on the table, it would roll around to its only stable face.

It turns out that this is indeed possible and one solution, consisting of a polyhedron with nineteen faces, is given in Conway and Guy (1969). A physical model of this, in solid brass, is shown in plate 29. From personal discussions with John Conway, the authors believe that this is unique, and we shall now sketch exactly how it is constructed.

We begin the construction with a prism such as that shown in figure 12.11. Each half of this prism is made up of m right-angled triangles, each of the angles at the centre is equal and is therefore $180°/m$. The hypotenuse of the largest triangle has length $r = r_0$ and is vertical when the solid is in equilibrium. The other hypotenuses are $r_n = \cos^n(180°/m)$ for $0 < n < m$, with $s = r_m = r\cos^m(180°/m)$ being collinear with $r =$

r_0. The ends of this prism are truncated obliquely as shown in figure 12.12, with two planar faces. We need to choose m, b and a sufficiently large so that the centre of mass lies below O but above P. The solid will then stand on only one lateral face, and on neither of the oblique ones. An involved calculation, in Conway and Guy (1969), shows that for $m = 9$ it is possible to achieve this provided that b is sufficiently large compared with a.

Chapter 13

FINDING SOME EQUILIBRIUM

If you throw a silver dollar down upon the ground, it will roll because it's round.

Attributed to Jack Palmer and Clarke van Ness (1939)

The words of the song are true enough though eventually the coin will lose its rotational energy, fall to one side and stop. It rolls, as does any circular disc, cylinder or ball, because its centre of gravity remains a constant height above a horizontal surface independent of position.

Following on from the last chapter, which was all about balance, we consider a number of related rolling problems. The first involves two joined cones which appear to roll up a slope. Secondly we show a way of joining two identical discs so that they will roll with no preferred position. We then do something similar for ellipses, before finishing with a solid model based on a generalized ellipse, known as the 'super-egg'.

13.1 Rolling Uphill

We shall begin with an experiment involving two rolling cones. Cones are a basic shape in solid geometry with which we hope you are familiar. Figure 13.1 shows two cones joined at their bases. The experiment is conducted by resting the cones on supports at the bottom of a slope. However, these supports are not parallel, and they can be arranged in such a way that when released, the roller appears to move *uphill*. The reason for this is simple. As the supports on which the roller rests are not parallel, the centre of mass of the cones actually descends, although the position on which the roller rests does rise. This illustrates an

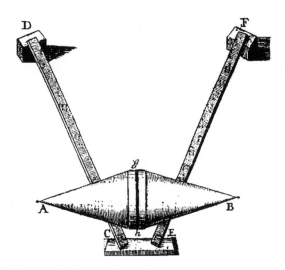

Figure 13.1. A roller that moves up a slope.

important principle: centre of mass will move downwards when-ever possible.

Figure 13.1 is taken from Leybourn (1694), in which the author claims that the rollers are a novel invention. There are various ways to make a modern version. Two traffic cones together with scaffold planks make a slightly larger and equally compelling practical experiment.

The uphill roller is a three-dimensional problem, so in order to describe it we need to sketch the situation carefully. This has been done from three perspectives in figure 13.2. We take the origin O to be where the two support lines meet, the angles these support lines make with the horizontal as t and with each other as $2s$. The centre of mass of the double cone, which has coordinates x and y, is labelled G, and the cones have a base radius of r and a height of h. From this it is clear in the end view that $\tan(u) = r/h$. There is a natural symmetry in the problem, so that we only mark one point of contact of the cone with its support, which we call A.

Our task is to describe the height y in terms of x, and hence derive criteria for y to decrease as x increases—so that G goes down even if A goes up.

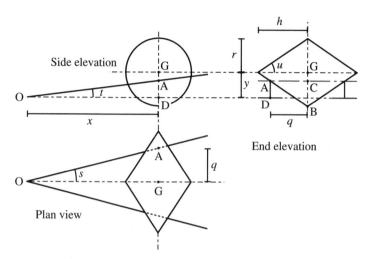

Figure 13.2. A sketch of the uphill roller.

From the plan view we have that

$$q = x \tan(s).$$

From the end view,

$$BC = q \tan(u) = x \tan(s) \tan(u).$$

Using the front and end views we have that

$$\begin{aligned}
y &= AD + CG \\
&= AD + BG - BC \\
&= x \tan(t) + r - x \tan(s) \tan(u) \\
&= r + x(\tan(t) - \tan(s) \tan(u)).
\end{aligned}$$

From this it is clear that if $\tan(t) - \tan(s) \tan(u) = \tan(t) - (r/h) \tan(s)$ is negative, then the roller will move uphill.

13.2 Perpendicular Rolling Discs

Let us continue with another practical experiment in which we take two identical thin, uniform discs. The experiment involves slotting the two identical discs together. The most important thing from a practical point of view is to ensure that the discs are fixed together radially and are perpendicular. We shall assume, for the sake of argument, that the radius of both discs is one unit and we shall ensure that the distance between the centres of the discs is exactly $\sqrt{2}$ units, which in practice is about 1.414.

You should now have two discs that look like those shown in plate 30. Now try rolling them along a horizontal table surface. They move in an intriguing way and do not seem to have a preference for any one orientation. In fact, in this section we will show that no matter what the orientation of this compound object, the centre of mass is indeed a constant distance from the table. Hence, the shape will roll along the table top without naturally moving towards, or resting in, one particular orientation.

What is particularly striking about the problem is that it is *fundamentally three dimensional*. Some problems, such as that of stacking dominoes, can easily be reduced to a two-dimensional problem that is much easier to solve. If you try to make these discs and roll them along a flat surface it will soon become apparent that there is a complex three-dimensional motion in play. Furthermore, the object we have made only makes contact with the surface of the table in *two* places, whereas when we talk about an equilibrium position of a solid on a plane we usually need at least *three* points of contact. Finally, there are obvious symmetries in the shape and in its motion. All these factors make modelling and describing the motion of this model quite an interesting problem.

Let us begin by describing the problem in more detail. We have two thin uniform discs of radius 1, called D_1 and D_2, with respective centres O_1 and O_2. These are slotted together along diameters with the centres a distance x apart, and the planes of the discs are perpendicular. We assume that this object rests on a horizontal plane surface. The disc D_1 will touch the table at a point C_1 on its edge. The angle between O_1O_2 and O_1C_1 is called t_1.

By symmetry, the centre of mass of the object lies halfway along the line connecting O_1 and O_2. Our task is to choose x such that, no matter the orientation of the object, the centre of mass lies a constant distance above the surface of the table. Of

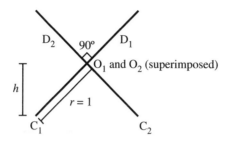

Figure 13.3. End view along $O_1 O_2$.

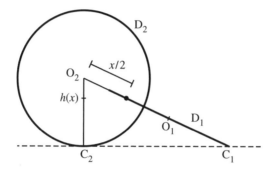

Figure 13.4. Side view when D_2 is vertical.

course, it is by no means obvious that such a choice is possible. Our first task is to consider some simple cases that will help us establish what the value of x might be.

There are two 'extreme' positions for the object. The first occurs when t_1 is $90°$, at which, by symmetry, the point of contact of D_2 with the table is also at $t_2 = 90°$. In this configuration the line connecting O_1 and O_2 lies above, but parallel to, the plane of the table itself. Hence x does not affect the height of the centre of mass above the table. In this configuration, looking along the line connecting O_1 and O_2 we see what is shown in figure 13.3.

From this it is clear that $h(x)$, the height of the centre of mass above the table, will be $\frac{1}{2}\sqrt{2}$. Since this does not depend on x, if h is to be kept constant then it must take this value.

Roll the discs to find the second 'extreme' position, which occurs when $t_1 = 0°$, forcing D_2 to lie in a vertical plane. Looking from the side we see what is shown in figure 13.4.

The distance from O_1 to C_1 is 1, and from O_1 to O_2 is x. Furthermore, the triangle $O_2C_2C_1$ forms a right-angled triangle. A simple application of similar triangles shows that $h(x)$ is given by the equation

$$h(x) = 1 - \frac{x}{2}\frac{1}{1+x} = \frac{2+x}{2+2x}.$$

Since we are trying to find a configuration in which the centre of mass lies a constant distance above the table, we are trying to solve

$$h(x) = \frac{2+x}{2+2x} = \frac{\sqrt{2}}{2}.$$

Rearranging this equation to solve for x gives

$$x = \sqrt{2}.$$

If the distance between O_1 and O_2 is chosen to be $\sqrt{2}$, then the centre of mass will be a distance $\frac{1}{2}\sqrt{2}$ above the table in both of the extreme positions. Physically, to achieve this we cut a slot $1 - \frac{1}{2}\sqrt{2} \approx 0.292$ times the radius in each disc.

The height of the centre of mass is now guaranteed to be the same in two positions, and more mathematical work using three-dimensional geometry is necessary to determine whether or not this height is totally independent of the position of the discs.

An alternative experimental approach is to get two discs whose thickness is small in relation to their radii and see what happens. Our first trial was with two plastic discs of 20 mm radius and 1 mm thickness, which used to come with packets of potato crisps. They rolled beautifully and provided excellent experimental evidence.

In the elementary geometric analysis of figures 13.3 and 13.4 we assumed that the discs were thin, i.e. that they had no thickness whatsoever. Further thought shows that this need not have been assumed in the first place provided that the *rims* of the discs are thin: nothing has been assumed about the rest of the discs. In fact, discs of almost any thickness can be used provided

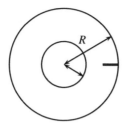

Figure 13.5. Slotting rings.

the perimeters are chamfered to a knife edge and that they are
symmetrical about their centre planes. Figure 13.4 shows that
when D_1 is at its closest to being horizontal it makes an angle of

$$t = \sin^{-1}\left(\frac{1}{1 + \sqrt{2}}\right) \approx 24.5°$$

to the horizontal: a chamfer at a sharper angle satisfies the condition that, as far as rolling is concerned, the discs are thin.

Unlike the solids of constant width described in section 10.3,
there is some merit in using a dense material to make this model.
The extra mass helps the discs roll along smoothly and overcome
air resistance as well as providing stronger support around the
slots.

The discs need not be complete circles and they make attractive toys. One possibility is to use the rings shown in figure 13.5,
where the inner radius is $(\sqrt{2} - 1)R$.

For the minimalist they need not involve discs or rings at all,
just a wire skeleton. Indeed, the outline of the shapes shown
in plate 31 is sufficient. This form is difficult to make and to
provide with strong joints. It does suggest that we are dealing
with the skeleton of a solid the surface of which can be created
from a flat sheet without any need to crease or stretch it in any
way. The outline, or pattern, can be found mathematically or
by experiment. Wipe some lipstick or engineer's blue round the
rims of the model, but only where they touch the surface when
rolling otherwise it can be a messy job. Roll the model over a
sheet of paper or card and there is the pattern ready to cut out.
As with all solid models made from card it is difficult to glue
the edges together, but it is worth a try. Failing this, the pattern

can be transferred to cloth, making allowances for the hems. With suitable stuffing you can have a soft toy, although now it is hardly a mathematical model!

In the rest of this section we shall return to our mathematical analysis of the rolling discs to show in detail that the height of the centre of mass does indeed remain a constant distance above the table. In the next section we shall do something very similar with two ellipses.

In order to consider the general situation we shall consider the contact point of D_1 with the table: we label this point of contact C_1. We assume that the object is sitting on the table, with D_2 touching at the point, not shown, that we call C_2. Ignoring the symmetry of D_1 and D_2, the position of C_1 is enough to dictate the position of the object uniquely.

To help us consider the position of the point C_1, we shall draw a line through the centre of each of the discs. Note that this line also lies in the plane of the discs. To describe the position of C_1 we measure an angle from the line O_1O_2 to the line O_1C_1 in the plane of D_1, and we refer to this angle as t_1. This situation is shown in figure 13.6, to which we shall refer in the forthcoming discussion. Note that the point C_2 is not shown, since this will be on the disc D_2, shown as a line in figure 13.6, and its position will depend on the angle t_1. We only need to consider values for angle t_1 that lie between $0°$ and $90°$, since the other situations are dealt with by symmetry.

We wish to mathematically describe the object relative to the table, or equally the table relative to the object. Having obtained our mathematical description, we will use this to show that the distance of the centre of mass of the object from the table is constant, regardless of the angle t_1, and hence the position of the point of contact, C_1.

One way to describe a plane mathematically is to specify three points in space. However, we only have *two* points of contact, C_1 and C_2, with the plane of the table—insufficient to describe the plane of the table. Another way to describe a plane is to find a line in the plane and another point that is not on that line. To use this second idea we need to take advantage of the following

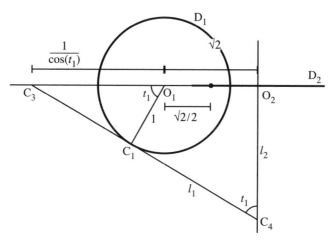

Figure 13.6. Looking in the plane of disc D_1.

observation: the tangent line to D_1 at the point C_1 lies in the plane of the table. This line is labelled in figure 13.6 as l_1.

By considering right-angled triangles we can also see immediately that the distance from O_1 to C_3 is $1/\cos(t_1)$. Note that since l_1 lies in the plane of the table, C_3 is the intersection of the line joining the centres of the discs and the plane of the table itself.

The next line we consider is that perpendicular to the plane of D_2 through the centre of D_2. We have labelled this line l_2. Since this line is also in the plane of D_1, it intersects the table where l_1 and l_2 meet, a point we label as C_4. Again, simple considerations show that l_1 and l_2 intersect at an angle of t_1, and furthermore that the distance from C_4 to O_2 is given by

$$\frac{1}{\tan(t_1)}\left(\frac{1}{\cos(t_1)} + \sqrt{2}\right) = \frac{1 + \sqrt{2}\cos(t_1)}{\sin(t_1)}.$$

Note that we have already considered the case when $t_1 = 0°$, in which l_1 and l_2 do not intersect.

Next we have to consider a different plane altogether: that which is defined by the three points O_2, C_2 and C_4. By considering these, we take account of the line l_1 and the point C_2, and these are sufficient to describe the plane of the table. These form a right-angled triangle at O_2, and since the line tangent to C_2 also

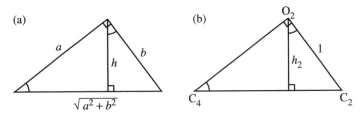

Figure 13.7. O_2C_4 is perpendicular to D_2;
C_2C_4 lies in the plane of the table.

lies in the plane of the table, the plane containing O_2, C_2 and C_4 is *perpendicular to the table itself.* Hence we can use this triangle to find the height, h_2, of O_2 from the table, as indicated by figure 13.7.

In part (a) of the figure we have the general situation. By considering similar triangles, and the Pythagorean theorem, we can easily calculate

$$h = \frac{ab}{\sqrt{a^2 + b^2}}.$$

Applying this to the situation we need to consider, shown in part (b), allows us to calculate h_2 using the length obtained above for the line C_4O_2. Note that a lengthy, but routine, calculation has been omitted here before we obtain

$$h_2 = \frac{\sqrt{2}\cos(t_1) + 1}{\cos(t_1) + \sqrt{2}}.$$

Thus we are almost there, but actually we did not want h_2 but the height of the centre of mass from the table. To calculate this we consider a third plane. This is a vertical plane passing through O_1 and O_2. We already have a number of points in this plane in our previous diagrams, and combining this information we arrive at figure 13.8.

We are trying to find h, which is easily calculated as the average of h_1 and h_2. Again using similar triangles we see that

$$h_1 = \frac{h_2}{\sqrt{2}\cos(t_1) + 1} = \frac{1}{\cos(t_1) + \sqrt{2}},$$

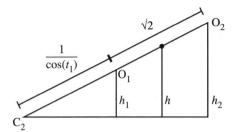

Figure 13.8. Calculating h, given t_1.

and so

$$h_1 + h_2 = \frac{1}{\cos(t_1) + \sqrt{2}} + \frac{\sqrt{2}\cos(t_1) + 1}{\cos(t_1) + \sqrt{2}}$$
$$= \frac{\sqrt{2}\cos(t_1) + 2}{\cos(t_1) + \sqrt{2}} = \sqrt{2}.$$

The last equality shows that $h_1 + h_2$ does not depend on t_1 after all, and that the value of h is constant at $\frac{1}{2}\sqrt{2}$, as claimed.

13.3 Ellipses

Although the above construction was actually very simple, the calculations provided which show that the construction really worked were quite involved. However, we can actually make this construction more general still and slot together identical ellipses in a similar fashion. Just as the word 'circle' refers to the boundary of a disc, so strictly speaking we need planar shapes with elliptical boundaries—we simply refer to them as 'ellipses' since the abuse should cause no confusion. Look ahead to plate 32 in which models are shown. However, before we can make such a model we will need to cut out ellipses. In light of section 1.4, some of you may be wondering how this can actually be done. One method, using an arrangement of linkages that assembles some of the components we have already described in chapter 2, is given by Yates (1938). Our goal is more modest, and we will be satisfied if we can draw an ellipse. Hence, we will remind you of some relevant properties of the ellipse and in doing so we will explain how to draw them.

An ellipse is generated by taking a slice through a cone in such a way that we produce a closed curve, much like a squashed circle. The usual form of the equation for an ellipse is

$$\frac{x^2}{a^2} + \frac{y^2}{b^2} = 1, \tag{13.1}$$

where a and b are positive constants. If b equals a then (13.1) simply reduces to $x^2 + y^2 = a^2$, which is the equation of the circle. It is a usual convention to assume that $a > b$ and then refer to the longer axis as the *major axis* and the shorter as the *minor axis*. Two other points of importance on the ellipse are known as the foci. These lie on the major axis at the coordinates $(\pm\sqrt{a^2 - b^2}, 0)$. While this looks complicated, actually locating the foci is simple in practice. Open compasses to half the length of the major axis. Place them on the intersection of the minor axis and the ellipse. The points at which these lines intersect with the major axis are the foci, which we shall refer to as points F and F'.

The foci are the key to many of the interesting properties of the ellipse, which we shall summarize now. Let us take any point P on the ellipse. Imagine a line from F to P and back to F'. This is shown in figure 13.9. The first important property is that the length of the line FPF' is independent of the position of P. This gives us a first method of drawing an ellipse. Take two pins, a loop of string and a pencil. Place the two pins at the foci and keeping the string under constant tension we can draw an ellipse with the end of the pencil. This is actually rather hard to do, since the string slips off the pencil and the inevitable elasticity in the string causes observable errors. Even if the method could be perfected from a practical point of view, those who have followed chapter 1 should be suspicious since the pencil and pins have circular and not necessarily identical cross-section diameters and the string is broad. Hence we are not really at the foci, and are not really following the centre of the piece of string. If the string is inelastic, and the pins and pen have identical diameters, then in fact a true ellipse is drawn. But since the method is so frustrating anyway we shall not examine any theoretical errors that arise.

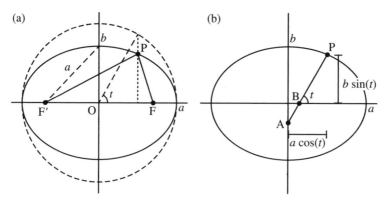

Figure 13.9. Important properties of an ellipse.

The next interesting property is that the normal to the ellipse at the point P is the bisector of the angle FPF′. This means, from a physical point of view, that an elliptical mirror will reflect light emitted from one focus back to the other—hence the origin of the term, perhaps. One way to find the tangent to the ellipse is to construct a line perpendicular to the normal. We take a more direct approach to give the explicit equation for the line tangent to the ellipse at the point (x', y'). Since (13.1) holds, we have the gradient given using calculus as

$$y \frac{dy}{dx} = -\frac{b^2}{a^2} x.$$

From this it is possible to calculate that the tangent line at (x', y') is

$$\frac{xx'}{a^2} + \frac{yy'}{b^2} = 1. \tag{13.2}$$

Although this does not look much like the equation of a straight line, compare this with (10.1).

Another characterization of the ellipse is to take a parameter t to be any real number and consider (x, y) defined to be

$$(x, y) = (a \cos(t), b \sin(t)). \tag{13.3}$$

Substituting these values for x and y into the left-hand side of (13.1) reduces to

$$\cos^2(t) + \sin^2(t),$$

which equals 1 by the Pythagorean theorem. That is to say, the points given by (13.3) all lie on the ellipse (13.1). A circle of radius a drawn around the ellipse is often called the *auxiliary circle*. The parameter t is then the angle shown in figure 13.9 and is *not* the angle between OP and the major axis.

A second practical method for drawing an ellipse is sometimes referred to as the *trammel of Archimedes*. Take a line AP of length a and mark on the position B so that PB $= b$. Then constrain this line to move so that A is on the y-axis and B is on the x-axis. From figure 13.9(b) it is clear that the x-coordinate of the point P is given as $a\cos(t)$ and the y-coordinate is simply $b\sin(t)$, proving that the equation for P is exactly (13.3) and hence lies on the ellipse. The trammel is a reasonably practical device, although care needs to be taken in making it so that neither friction nor play cause problems. You might like to consider what shape is drawn if the guides in the trammel are not quite at right angles to each other. An alternative arrangement is used by the commercial ellipsograph shown in figure 13.10. In this arrangement point B is constrained to move on the major axis by a slider. The slider on the minor axis has been replaced by an ingenious Scott Russell linkage of exactly the type described in section 2.4. This constrains the point A in figure 13.10(a) to move in a vertical straight line. In this ellipsograph only half an ellipse can be drawn at once, and the device must be spun on its pointer to complete the curve. However, there is no trammel to obstruct the curve, making smaller ellipses a practical proposition. We have seen one alternative batch of Omicron model 17 ellipsographs in which the short straight guide in the Scott Russell linkage has been replaced by a single link. This is in fact exactly the grasshopper mechanism we already described in section 2.4. However, now the point A does not move on a straight line but on an arc of a large circle. As a result the path of the pencil is not a true ellipse but only an approximate ellipse.

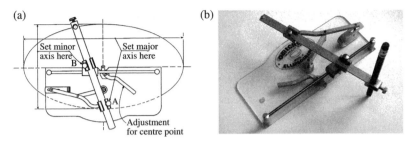

Figure 13.10. The Omicron model 17 ellipsograph.

13.4 Slotted Ellipses

Ellipses can be used in place of circular discs and, if positioned correctly, these roll equally well. Take two copies of the ellipse described by (13.1), and slot them together along the x-axis. If we repeat the analysis we undertook for the discs, then it is clear that the centre of mass will be a distance $\frac{1}{2}\sqrt{2}b$ above the table, regardless of the distance x. In the other extreme position an analogous analysis involving the tangent to the ellipse given by (13.2) can be used to show that the only possible value of x that also gives this result is

$$x = \sqrt{4a^2 - 2b^2}. \tag{13.4}$$

We have omitted details of these calculations and those that prove that using this configuration the height is maintained in all orientations of the ellipses.

Notice that we have not assumed that the x-axis is the major axis, although implicit in (13.4) is the assumption that

$$\sqrt{2}a > b.$$

If $a > b$, so that the x-axis is the major axis, then this is satisfied. However, one particularly interesting case occurs when $x = 0$, which is to say when there is no displacement between the centres of the ellipses. This corresponds to $b = \sqrt{2}a$, and a model of this is shown at the top of plate 32. It exhibits a most intriguing rolling action that static pictures cannot hope to capture.

Its behaviour might have been expected though, since any slice through a circular cylinder is elliptical and the cylinder itself can roll. The particular ratio of a to b is such that the ellipses can be slotted together at right angles, and consequently rolling in two perpendicular directions is possible. The geometry is quite different from that of the previous examples, where the ability to roll smoothly was not so easily predicted.

The plywood models in plate 32 were made with $b = 0.8$ to give a reasonable depth of slot for each model. The most difficult part of making the models is cutting the slots. A saw is the obvious tool to use and the way the elliptical discs were made was to cut the slots first and then paste over a photocopy of the ellipses and finally finish to this shape. It is not worth expending effort and time cutting out two ellipses before slotting because success then depends entirely on the simple saw cuts in each disc. Start with wood that is slightly oversize, cut the slots overdeep and then, holding the wood to the light, paste on the patterns. This works well and rounding the edges improves the rolling performance.

13.5   The Super-Egg

We shall end this chapter with one last gravity-defying model. This is a direct generalization of the ellipse, where we have taken not the equation (13.1) but rather a generalization

$$\left| \frac{x}{a} \right|^n + \left| \frac{y}{b} \right|^n = 1, \tag{13.5}$$

where $n > 2$. To make an 'egg' rather than a flat object we shall form a solid of revolution about the vertical axis. That is to say, using the y-axis as an axle we rotate the egg out of the x, y-plane. For the ellipse this solid is known as an *ellipsoid* and when (13.5) is used we refer to the solid as a *super-egg*. To illustrate this discussion we take $n = 2.5$, $a = 1$ and $b = 2$. The points with $y > 0$ that satisfy (13.5) are plotted as a solid line together with an ellipse, as a dashed line, for comparison in figure 13.11. Hardwood models of the corresponding super-egg are shown in

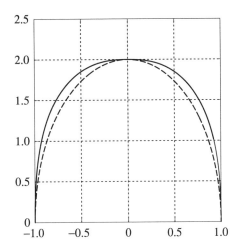

Figure 13.11. A super-egg.

plate 33. The larger egg to the left is made from the wood of the tulip tree, *Liriodendron*.

For the purposes of our analysis we shall exploit this axis of symmetry, so it will be sufficient to consider the cross-section in only two dimensions.

Imagine placing an ellipse (or ellipsoid) on a horizontal table with the major axis also horizontal. It is intuitively clear that if the ellipse is tipped from this position then the centre of mass will rise, so this position is stable. Similarly, if the ellipse is moved so that the minor axis is now horizontal, then in theory at least there is a point of equilibrium. This will be unstable in the sense that any displacement, however small, will move the centre of mass closer to the table.

We can formalize this by applying the Pythagorean theorem to points that satisfy (13.1). It is easy to see that if r is the distance of the points from the origin then

$$r^2 = b^2 + \left(1 - \frac{b^2}{a^2}\right)x^2.$$

Initially, if we have the resulting ellipse resting on the horizontal plane $y = -b$, then whether r increases or decreases for small x is determined by whether $b/a > 1$, which is equivalent to whether $b > a$. If $a > b$, which is the usual situation, then we

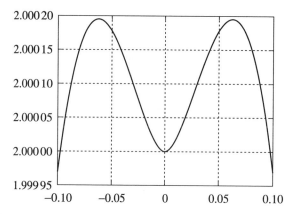

Figure 13.12. The distance from points on (13.5) to
the origin for $n = 2.5$, $a = 1$, $b = 2$.

have the major axis horizontal. It follows that $(1 - (b^2/a^2)) > 0$
and so r increases with x. The centre of mass is lifted, so the
position is stable. Conversely, if $b > a$, then we have the minor
axis horizontal and it follows that $(1 - (b^2/a^2)) < 0$ and so r
decreases with x. The centre of mass falls giving an unstable
equilibrium.

What about the super-egg? If we apply the Pythagorean theorem one last time to points that satisfy (13.5), then we are left
to consider

$$r^2 = x^2 + \frac{b^2}{a^2}(a^n - x^n)^{2/n}.$$

The only case of interest is when $b > a$, and here it can be shown
in general that the centre of mass *rises initially*. We shall give
one numerical example with $n = 2.5$, $a = 1$ and $b = 2$. In this
we have taken points (x, y) on (13.5) and plotted the distance
from the origin against the x-coordinate for small values of x
in figure 13.12. Notice the exaggerated vertical axis, which indicates that the distance from the origin hardly changes at all, but
more importantly it is clear that initially the distance increases.
If the egg is displaced slightly from the upright position when
the point of contact with the table is $x = 0$, then the centre
of mass will fall by moving back towards this equilibrium. This

well of stability allows the super-egg to apparently defy gravity by standing on its end.

Although an explanation of the result is beyond the scope of this book, try this final experiment. With your super-egg lying with its major axis horizontal on a flat table give it a good spin and watch its behaviour very carefully.

Epilogue

In this book we have tried to illustrate why mathematicians should take the practical problems of engineering seriously. We have also tried to illustrate some mathematics by suggesting experiments and the construction of mathematical models.

We have not provided a systematic treatment of the Platonic solids: this is a topic that is so comprehensively covered in Cundy and Rollet (1961), and in many other books, that we have little if anything new to add. Similarly, we do not include anything about sundials, which are an excellent source of basic mathematical activities. However, this is extensive enough a field to deserve the sort of dedicated treatment found in the excellent books of, for example, Rohr (1970) and Waugh (1973). Those who have enjoyed reading this book might like to consider sundials as their next fruitful source of mathematical construction projects. For a more mathematical treatment you might appreciate working through Yates (1949), which is a carefully structured set of exercises in geometry that have much in common with the topics we have covered.

We have also tried to illustrate what happens to familiar and apparently simple things when you dig a little deeper, or change something slightly. Neither of us realized quite how interested we would become in these topics when we embarked on this book. We can only encourage you to continue where we have stopped.

References

Adleman, L. M. 1994. Molecular computation of solutions to combinatorial problems. *Science* 266:1021–24.

Artobolevsky, I. I. 1976. *Mechanisms in Modern Engineering Design.* Moscow: MIR.

Ashton, A. 1999. *Harmonograph: A Visual Introduction to Harmony.* Presteigne: Wooden Books.

Beiler, A. H. 1964. *Recreations in the Theory of Numbers.* New York: Dover.

Bird, J. 1767. *The Method of Dividing Astronomical Instruments.* London: John Nourse.

Bolt, A. B., and J. E. Hiscocks. 1970. *Machines, Mechanisms and Mathematics.* London: Chatto and Windus.

Bond, G. E. 1887. *Standards of Length and Their Practical Application.* East Hartford, CT: Pratt and Whitney.

Bourne, J. 1846. *A Treatise on the Steam Engine in Its Application to Mines, Mills, Steam Navigation and Railways.* Harlow: Longman.

Branfield, J. R. 1969. An investigation. *The Mathematical Gazette* 53(385):240–47.

Brunton, J. 1973. The cube dissected into three yángmǎ. *The Mathematical Gazette* 57(399):66–67.

Cadwell, J. H. 1966. *Topics in Recreational Mathematics.* Cambridge University Press.

Cajori, F. 1909. *A History of the Logarithmic Slide Rule.* New York: J. F. Tapley. (Note: this book contains a flawed conclusion as to the priority of the invention of the sliderule (see Cajori 1920).)

———. 1916. *William Oughtred: A Great Seventeenth-Century Teacher of Mathematics.* Chicago, IL: Open Court.

———. 1920. On the history of Gunter's scale and the slide rule during the seventeenth century. *University of California Publications in Mathematics* 1(9):187–209.

Casselman, B. 2000. Pictures and proofs. *Notices of the American Mathematical Society* 47(10):1257-66.

Chapman, A. 1990. *Dividing the Circle. The Development of Critical Angular Measurement in Astronomy 1500-1850.* Chichester: Ellis Horwood.

Chaucer, G. 1872. *A Treatise on the Astrolabe.* Extra Series, volume 16. London: Early English Text Society. (Originally published in 1391.)

Cole, D. E. 1972. The Wankle engine. *Scientific American* 227(2):14-23.

Conway, J. H., and R. K. Guy. 1969. Problem 66-12, stability of polyhedra. *SIAM Review* 11(1):78-82.

Cotter, C. H. 1981. Edmund Gunter (1581-1626). *Journal of Navigation* 34(3):363-67.

Crathorne, A. R. 1908. The Prytz planimeter. *The American Mathematical Monthly* 15(3):55-57.

Cundy, H. M., and A. P. Rollet. 1961. *Mathematical Models.* Oxford University Press.

Derham, W. 1717. Extracts from Mr Gascoigne's and Mr Crabtree's letters, proving Mr Gascoigne to have been the inventor of the telescopick sights of mathematical instruments, and not the French. *Philosophical Transactions* 30:603-10.

Dickinson, H. W. 1963. *A Short History of the Steam Engine*, 2nd edn. London: Frank Cass. (Originally published in 1938.)

Dobson, A. B. 1972. *The Slide-Rule in Everyday Use.* Weymouth: Blundell Harling.

Dudeney, H. E. 1907. *The Canterbury Puzzles.* London: Thomas Nelson.

Dunkerley, S. 1912. *Mechanism.* Harlow: Longman.

Edwards, C. H. 1979. *The Historical Development of the Calculus.* Springer.

Eisner, L. 1959. Leaning tower of the physical reviews. *American Journal of Physics* 27(2):121-22.

Emmerson, G. S. 1977. *John Scott Russell.* London: John Murray.

Euclid. 1956. *The Thirteen books of Euclid's Elements*, with introduction and commentary by Sir Thomas Heath. London: Dover.

Euler, L. 1778. De curvis triangularibus. *Acta Academiae Scientiarum Imperialis Petropolitanae* 2:3-30.

Euler, L. 2000. *Foundations of Differential Calculus.* Springer. (Translated from the Latin *Institutiones Calculi Differentialis* (1755) by J. Blanton.)

Fahmi, N. 1965. On note 3022—trisection of the angle. *The Mathematical Gazette* 49(367):80–81.

Farthing, S. 1985. Theory of the hatchet planimeter. *Sādhanaā* 8(4): 351–59.

Flamsteed, J. 1725. *Historia Coelestis Britannica.* London: H. Meere.

Foote, R. L. 1998. Geometry of the Prytz planimeter. *Reports on Mathematical Physics* 42:249–71.

Frederickson, G. N. 1997. *Dissections: Plane and Fancy.* Cambridge University Press.

——. 2002. *Hinged Dissections: Swinging and Twisting.* Cambridge University Press.

Gardner, M. 1966. *New Mathematical Diversions.* New York: Simon and Schuster.

Gatterdam, R. W. 1981. The planimeter as an example of Green's theorem. *American Mathematical Monthly* 88(9):701–4.

Goho, K., M. Kimiyuki and A. Hayashi. 1999. Development of a roundness profile measurement system for parallel rollers based on a v-block method. *International Journal of the Japanese Society of Mechanical Engineers* C42(2):410–15.

Grant, G. B. 1907. *A Treatise on Gear Wheels,* 10th edn. Boston, MA: Grant Gear Works.

Gray, C. G. 1972. Solids of constant breadth. *The Mathematical Gazette* 56(398):289–92.

Hall, J. F. 2005. Fun with stacking blocks. *American Journal of Physics* 73(12):1107–16.

Hardy, G. H. 1967. *A Mathematician's Apology.* Cambridge University Press.

Hartree, D. R. 1938. The mechanical integration of differential equations. *The Mathematical Gazette* 22(251):342–65.

Hayes, G. 1990. *A Guide to Stationary Steam Engines.* Ashbourne: Moorland.

Heather, J. F. 1871. *A Treatise on Mathematical Instruments,* 10th edn. London: John Weale.

Hele Shaw, H. S. 1885. The theory of continuous calculating machines and a mechanism of this class on a new principle. *Philosophical Transactions of the Royal Society of London* 176(2):367–402.

Henrici, O. 1894. Report on planimeters. *British Association for the Advancement of Science, Report of the 64th Meeting*, pp. 496–523.

Heppes, A. 1967. A double tipping tetrahedron. *SIAM Reviews* 9:599–600.

Hoelscher, R. P., J. N. Arnold and S. H. Pierce. 1952. *Graphic Aids in Engineering Computation.* New York: McGraw-Hill.

Holmes, W. T. 1978. *Plane geometry of rotors in pumps and gears.* Manchester: The Scientific Publishing Company.

Hopp, P. 1998. *Slide Rules, Their History, Models and Makers.* Mendham, NJ: Astragal.

Horsley, S. (ed.). 1782. *Isaaci Newtoni Opera quae extant Omnia.* London: Joannes Nichols.

Kamenetskiĭ, I. M. 1947. Rešenie geometriceskoi zadači L. Lyusternika. ('Solution of a geometrical problem of L. Lyusternik.') *Uspekhi Matematicheskikh Nauk* 2(2):199–202.

Keady, G., P. J. Scales and G. F. Fitz-Gerald. 2000. Envelopes generated by circular-arc coupler-bars in James Watt's four-bar linkages. *Proceedings of the Engineering Mathematics and Applications Conference*, pp. 171–74. Melbourne: RMIT.

Kearsley, M. J. 1952. Curves of constant diameter. *The Mathematical Gazette* 36(317):176–79.

Keay, J. 2000. *The Great Arc.* London: Harper Collins.

Kempe, A. B. 1875. On a general method of producing exact rectilinear motion by linkwork. *Proceedings of the Royal Society of London* 23(163):565–77.

——. 1876. On a general method of describing plane curves of the nth degree by linkwork. *Proceedings of the London Mathematical Society* 7:213–16.

——. 1877. *How to Draw a Straight Line: A Lecture on Linkages.* London: Macmillian.

Leybourn, W. 1694. *Pleasure with Profit: Consisting of Recreations of Diverse Kinds.* London: Richard Baldwin and John Dunton.

Littlewood, J. E. 1986. *Littlewood's Miscellany.* Cambridge University Press.

Loomis, E. 1968. *The Pythagorean Proposition.* Washington, DC: National Council of Teachers of Mathematics.

Ludlam, W. 1786. *An Introduction and Notes on Mr Bird's Method of Dividing Astronomical Instruments.* London: John Sewell.

Lyusternik, L. A. 1964. *Convex Figures and Polyhedra.* London: Dover.

MacHuisdean, H. 1937. *The Great Law*. Erlestoke: Erlestoke Press.

Maxwell, E. A. 1954. *An Analytical Calculus for Schools and Universities*. Cambridge University Press.

Meskens, A. 1997. Michiel Coignet's contribution to the development of the sector. *Annals of Science* 54:143–60.

Mills, B. D. 1958. The Nelson slide rule. *American Mathematical Monthly* 65(3):194–95.

Minkowski, H. 1904. Über die Körper konstanter Breite. *Recreational Mathematics, Moscow Mathematical Society* 25:505–8.

Nash, D. H. 1977. Rotary engine geometry. *Mathematics Magazine* 50(2):87–89.

Needham, T. 1997. *Visual Complex Analysis*. Oxford University Press.

Oughtred, W. 1634. *The Circles of Proportion and the Horizontal Instrument*. Oxford University Press.

Pedersen, O. 1987. The Prytz planimeter. In *From Ancient Omens to Statistical Mechanics, Essays on the Exact Sciences* (ed. J. L. Berggren and B. R. Goldstein). Acta Historica Scientiarum Natutalium et Medicinalium, volume 39, pp. 259–70. Copenhagen: University Library.

Reason, R. E. 1966. *Report on the Measurement of Roundness*. Leicester: Rank Taylor Hobson.

Rohr, R. R. J. 1970. *Sundials: History, Theory and Practice*. London: Dover.

Rouse Ball, W. W. 1960. *Mathematical Recreations and Essays*, 11th edn. London: Macmillan.

Sangwin, C. J. 2002. Newton's polynomial solver. *Journal of the Oughtred Society* 11(1):3–7.

Sangwin, G. W. F. 1963. The magnameta oil tonnage calculator. British Patent 919063.

Schwamb, P., and A. L. Merrill. 1984. *Elements of Mechanism*. Bradley, IL: Lindsay Publications. (First published 1904.)

Smith, G., L. Wood, M. Coupland and B. Stephenson. 1996. Constructing mathematical examinations to assess a range of knowledge and skills. *International Journal of Mathematics Education in Science and Technology* 27(1):65–77.

Snodgrass, B. 1959. *Teach Yourself the Slide Rule*. London: English Universities Press.

Snow, L. T. 1903. Planimeter. United States Patent 718166.

Stewart, I. 1989. *Galois Theory*, 2nd edn. Chapman & Hall.

Stone, E. 1743. *New Mathematical Dictionary*, 2nd edn. London: printed for W. Innys, T. Woodward, T. Longman and M. Senex.

——. 1753. *The Construction and Principal Uses of Mathematical Instruments Translated from the French of M. Bion, Chief Instrument Maker to the French King. To Which Are Added the Construction and Uses of Such Instruments as are Omitted by M. Bion, Particularly of Those Invented or Improved by the English*, 2nd edn. London: J. Richardson.

Sylvester, J. J. 1875a. On recent discoveries in mechanical conversion of motion. *Proceedings of the Royal Institution of Great Britain* 7: 179–98.

——. 1875b. The plagiograph and the skew pantigraph. *Nature* 12: 214–16.

Taylor, E. G. R. 1954. *The Mathematical Practitioners of Tudor and Stuart England*. Cambridge University Press (for the Institute of Navigation).

Turnbull, H. W. (ed.). 1959–61. *The Correspondence of Isaac Newton*. Cambridge University Press for The Royal Society.

Turner, A. J. 1973. Mathematical instruments and the education of gentlemen. *Annals of Science* 30:51–88.

——. 1981. William Oughtred, Richard Delamain and the horizontal instrument in seventeenth century England. *Annali Dell Intituto E Museo Di Storia Della Scienze Firenze* 6(2):99–125 (plus plates). (Note: the horizontal instrument is a form of sundial, not a slide rule.)

Watkins, G. M. 1953. The vertical winding engines of Durham. *Transactions of the Newcomen Society* 29:205–19.

Waugh, A. E. 1973. *Sundials: Their Theory and Construction*. London: Dover.

Whewell, W. 1819. *An Elementary Treatise on Mechanics: Designed for the Use of Students in the University*. Cambridge University Press.

Whiteside, D. T. 1961. Patterns of mathematical thought in the later seventeenth century. *Archive for the History of Exact Sciences* 1: 179–388.

Yates, R. C. 1938. A linkage for describing curves parallel to the ellipse. *American Mathematical Monthly* 45(9):607–8.

——. 1949. *Geometrical Tools: A Mathematical Sketch and Model Book*. Saint Louis, MO: Educational Publishers.

Index

Allen, Elias, 235
Amsler, Jacob, 158
analogue computer. *See* computer, analogue
angle, 87; bisection of, 100; trisection of, 72, 102, 105–8, 110; units of measurement, 92–95
arch, 131
Archimedes of Syracuse, 146; trammel of, 290
area, 53, 112, 118, 138–70, 194; of circle. *See* circle, area of; of parallelogram. *See* parallelogram, area of; of triangle. *See* triangle, area of; surface, 128
axioms, 70

Bacon, Roger, 183
bearing, 188; thrust, 128
Beghin, Auguste, 237
Bernoulli, lemniscate, 58
binary, 182–84
Bird, John, 94
Bissaker, Robert, 236
Board of Longitude, 94
Boulton, Matthew, 22
Brahe, Tycho, 90
bridge, Royal Albert, 132
Brunel, Isambard Kingdom, 39, 132

calendar, 182
cam, 15, 197–99
cantilever, 256
Cassini, Jacques, 67
catenary, 127, 130
centreless grinding, 204

Chaucer, Geoffrey, 89
Chebyshev, Pafnuty, 28, 41, 62
circle, 70, 144, 188, 190, 279; arc of, 5, 60; area of, 144, 239; auxiliary, 290; circumscribed, 205; inscribed, 205; inverse of, 35; least squares reference, 205; minimum zone, 205; of proportion, 234, 251–54; squaring of, 72
CNC milling, 12, 197, 203
coin, 188
complete parallel motion. *See* linkage, parallel
computer: algebra system, xvii; analogue, xv, 170, 227; parallel, 118
cone, 126, 278
constructible. *See* construction, geometric
construction: geometric, 69–73
Conway, John, 275
coupling, Oldham, 201
Crowther, Phineas, 24
cube, 113, 126
Cundy, H. Martyn, xvi
curvature: centre of, 13; radius of, 132, 195
curve: algebraic, 59; convex, 190; trancendental, 127
cylindricity, 205

Delamain, Richard, 234–36, 251–54
disc. *See* circle
dissection, 112–26; Dudeney, 8; Duijvestijn, 115; hinged, 7, 9; Willcocks, 115